Springer Series in Advanced Manufacturing

Other titles in this series

Assembly Line Design
B. Rekiek and A. Delchambre

Advances in Design
H.A. ElMaraghy and W.H. ElMaraghy (Eds.)

Effective Resource Management in Manufacturing Systems: Optimization Algorithms in Production Planning
M. Caramia and P. Dell'Olmo

Condition Monitoring and Control for Intelligent Manufacturing
L. Wang and R.X. Gao (Eds.)

Optimal Production Planning for PCB Assembly
W. Ho and P. Ji

Trends in Supply Chain Design and Management: Technologies and Methodologies
H. Jung, F.F. Chen and B. Jeong (Eds.)

Collaborative Product Design and Manufacturing Methodologies and Applications
W.D. Li, S.K. Ong, A.Y.C. Nee and C. McMahon (Eds.)

Process Planning and Scheduling for Distributed Manufacturing
L. Wang and W. Shen (Eds.)

Stanislav S. Makhanov
Weeracha i Anotaipaiboon

Advanced Numerical Methods to Optimize Cutting Operations of Five-Axis Milling Machines

With 113 Figures and 16 Tables

Springer

Stanislav S. Makhanov
School of Information and Computer Technology
Sirindhorn International Institute of Technology
Thammasat University
131 Moo 5, Tiwanont Road
Bangkadi, Pathum Thani 12000
Thailand
makhanov@siit.tu.ac.th
http://www.5axis-thai.com/

Weerachai Anotaipaiboon
National Science and Technology Development Agency
111 Thailand Science Park
Klong 1, Klong Luang
Pathumthani, 12120, Thailand
weerachai@nstda.or.th

Library of Congress Control Number: 2007922327

ISBN 978-3-540-71120-9 Springer Berlin Heidelberg New York

Springer is a part of Springer Science+Business Media

springer.com

Typesetting: Digital data supplied by author
Production: LE-TEX Jelonek, Schmidt & Vöckler GbR, Leipzig
Cover: eStudio Calamar S.L., F. Steinen-Broo, Girona, Spain

SPIN 11803164 89/3100/YL - 5 4 3 2 1 0 Printed on acid-free paper

To our families

Preface

Tool path generation and optimization for multi-axis milling machines is a new application area for computational mathematics requiring knowledge of numerical methods, differential and computational geometry, solid modeling, and optimization methods, as well as mechanical engineering and NC (Numerical Control) programming.

This book presents a computational framework for designing efficient computational algorithms for tool path optimization. We focus on five-axis machining, but the methods can also be applied to three- and four-axis machines.

The first three chapters of the book are tutorial, exposing readers to the basic knowledge necessary for five-axis cutting, such as G-code programming, differential geometry, and fundamental issues of tool path planning. These chapters can be used as an introduction to five-axis machining in the framework of an undergraduate or graduate course in computer aided manufacturing.

Advanced numerical methods introduced in the subsequent chapters include grid generation methods, space-filling curves, shortest path optimization, and the ramifications of each technique. This part of the book can be used for undergraduate and graduate research as well for corporate research and development. The authors have first-hand experience in designing, programming, and verifying five-axis optimization algorithms. Building on this experience, we present the numerical methods in detail and provide all of the basic computational formulas. Moreover, we illustrate and analyze the methods through numerous practical numerical examples, so that readers can understand, reproduce, and program any of them.

The book also includes a review of modern tool path optimization methods (130 references) and short historical notes on grid generation and space-filling curves.

We would like to thank Erik Bohez, Mud-Ameen Munlin, Bert Lauwers, and Than Lin for fruitful discussions and suggestions.

We also wish to acknowledge the National Science and Technology Agency, the National Electronic and Computer Technology Center, and the Thailand

Research Fund, whose sponsorship made it possible to conduct many machining experiments and measurements.

Bangkok,
December 2006

Stanislav Makhanov, Ph.D.
Weerachai Anotaipaiboon, Ph.D.

Contents

List of Figures

List of Tables

1

Introduction

1.1 Motivation and Structure of the Book

A variety of on-going research is focused on the development and analysis of methods to decrease the time required to progress from the computer modeling of the design surface to the machining while maintaining or improving the quality of the surface. One of the most important areas is tool path planning for numerical control (NC) machining. The main goal is obtaining the cutter location and orientation data that allow for an efficient surface milling within an allowed machining error.

Five-axis NC machines are becoming increasingly popular due to their ability to handle geometrically complex workpieces composed of raw material such as wood, wax, rubber, metal, stone, plastic, etc. Moreover, up-to-date five-axis NC machines are characterized by a high material removal rate and an efficient surface finish up.

Typically, manufacturing of the design surface by an NC machine comprises two stages, a rough cutting and a finish machining. During the rough cutting, the raw material is removed as fast as possible while ensuring no excessive cutting or gouging. During the finish machining, the tool is placed at the maximum contact with the surface to remove the remaining excess and create a well-finished and accurate surface. After finishing, the remaining scallops which are inevitably generated on the machined surface must be removed by manual surface grinding and polishing. The finish machining and manual polishing stages require as much as 75% of the total machining time. Besides, manual polishing is prone to error and undesirable irregularities.

Five-axis machining offers an improvement in efficiency of both the rough and finish machining stages over the three-axis counterpart. In five-axis machining, the tool orientation relative to the workpiece can be controlled by two additional degrees of freedom so as to achieve higher machining efficiency. With these advantages, a large number of tool path planning methods for five-axis machining has been developed and presented in the literature.

In chapter 1 we present the most popular CAD/CAM data formats and give a short literature survey on mathematical methods for optimization of five-axis machining. The survey has been focused on tool path interpolators, adaptable geometric patterns and methods for tool posture and gouging avoidance.

Chapter 2 exposes the readers to basic knowledge required to perform five-axis cutting. The G-code programming, examples of five-axis machining of simple shapes and verification of the cut using solid modeling software is presented and discussed. The chapter can be used as a short introduction into five-axis machining in the framework of an undergraduate course in Computer Aided Design (CAD) and Computer Aided Manufacturing (CAM).

Chapter 3 introduces theories required to embrace the concepts of the tool path optimization for five-axis machining. The chapter presents such fundamental issues as kinematics of the five-axis machines, part surface representation, machining strip, tool orientation and gouging avoidance as well as the forward step error. A variety of configurations of the five-axis machines is also discussed and analyzed. This chapter can also be used at an undergraduate or a postgraduate level for CAD/CAM related studies.

Chapters 4 and 5 present advanced optimization schemes based on the adaptable geometric patterns, namely, the space-filling curves (SFC) and adaptive curvilinear grids. The SFC tool path has a number of attractive features such as the possibility to locally adapt the curve in such a way that the cutting device travels along the optimal direction. In addition, the entire surface is cut in one path eliminating the need of tool retractions. The use of the concept of curvilinear grids allows to simultaneously adapt the points on the tool paths to create more efficient zigzag, spiral or even SFC structures. The combination of the SFC and grid generation allows for tool paths on surfaces with complex irregular boundaries, cuts off, pockets, islands, etc.

Three-axis machines are often thought of as three dimensional plotters. However the five-axis machine is more like a big bore machine. That is why changing from three-axis to five-axis programming is not an easy task. In particular, the idea of optimizing rotations may seem totally foreign from the viewpoint of the three-axis machining. Therefore, chapter 6 presents the theory and practice of optimization of rotations for five-axis machining. Several optimization algorithms based on the shortest path techniques are presented and discussed.

Chapter 7 presents a theoretical background developed to construct numerical algorithms to minimize kinematics error introduced by the initial setup of five-axis milling machines. The initial setup consists of the position and orientation of the workpiece with respect to the mounting table and, optionally, the machine's initial configuration. Given a set of cutter contact points and tool orientations, a least-squares optimization procedure finds the optimal setup parameters.

1.2 CAD/CAM Formats

Sculptured or free-form surfaces are widely used in today's manufacturing industries for a variety of applications such as the production of dies, molds, aerospace and automotive parts, etc. The surfaces are usually characterized by complex geometries and variable curvatures. A single surface is usually composed of patches represented mathematically by parametric forms such as the Bezier surfaces, B-splines and NURBS. A design and manufacture of the sculptured surface parts is an expensive and time-consuming process. First, a design surface is transformed into a computer model (possibly with the help of CAD programs). The computer model is then used by the CAM programs to generate commands to move the cutting tool of the machine. The resulting set of tool positions and orientations constitutes a tool path to machine the desired surface.

Every CAD or CAD/CAM software uses an internal format to represent and control the required part. When the geometrical data is transferred from a CAD system to a CAD or CAM system, a neutral format for the data transfer is used. One of the most popular is the IGES (Initial Graphics Exchange Specification) format (see the history of the IGES format in [40]). The IGES format supports the use of surfaces defined by NURBS (Non Uniform Rational B-Splines) or derivatives of these representation. A good book for a beginner in NURBS is [31]. An advanced reader could use [98] and [24]. There are also several free libraries designed to control and manipulate the NURBS, such as NURBS++ package [1] and NURBS Toolbox [2]. Some free IGES-file processing tools are collected at the NIST/IGES web page [3].

The STL files, originally employed by the so-called layered manufacturing technologies such as the rapid prototyping, are now becoming more and more popular due to the simplicity of describing the part surfaces. As opposed to the complex description of surfaces employed by the IGES format, the STL format defines the surface as a collection of triangles each described by the coordinates of its three corners and a normal-vector. This technology provides an important platform for CAD/CAM applications due to the existence of many robust triangulation algorithms. Besides, the surface models are often composed of many patches. Therefore, by tessellating the patches and creating groups of triangles one can use many well established methods for treating intersections, trimming, shading, hidden surface removal and gouge protection [36]. The machining is usually performed by creating contours obtained by slicing the STL-surface [85, 115]. Of course, slicing of the NURBS surfaces is also possible, but it requires much more sophisticated techniques such as [81]. The contours are then saved using the SLC format (see [4] for instance).

Other popular CAD/CAM formats include STEP (Standard for the Exchange of Product Data), DXF (The Drawing Exchange Format from AutoDesk) and many others. As a matter of fact, the difference in data formats has created a large software industry specializing in transferring, adapting and processing the CAD/CAM files. A table of compatibility of the CAD/CAM formats can be found, for instance, at [5] or [6].

1.3 Short Literature Survey

Optimization of tool paths for five-axis machining may include many features and multiple criteria such as the accuracy, the length of the tool path, the machining time, the size of the remaining scallops, etc. It may also include gouging avoidance, satisfying the machine axis limits, maximizing the volume of the removed material, reducing the tool wear. The optimization may also take into account the thermal characteristics of the cutting process, the tool bending, the vibrations and jacks, the workpiece positioning and many other parameters. The criteria could also include the configuration of the machine or specific parts of the machine as well as the design of the clamping device. Readers interested in citations before 1997 could use a fairly comprehensive survey by Dragomatz and Mann [25]. The survey presents a classification of research papers on three-axis and five-axis machining related to geometries of the tool paths and tool positioning. These categories include: 1) systems, 2) isoparametric paths, 3) non-isoparametric paths, 4) planar pocketing paths, 5) sculptured surface pocketing paths, 6) roughing paths, 7) tool positioning, 8) offset surface methods, 9) five-axis machining, 10) mesh models, 11) pixel and point models, 12) simulation and verification. Of course, the above groups overlap. Techniques involved in one group could be also involved in another group. For example, systems for tool path generation may include all of the above mentioned techniques. Roughing paths may be generated by the isoparametric or non-isoparametric schemes and so on.

Our survey is focused on five-axis machining. It also includes the three-axis methods but as long as they can be extended to the five-axis case. Besides, we confine ourselves by techniques designed for cutting the part surface by bottom-edge of the tool, e.g., flat-end milling and fillet milling. Many interesting methods designed for five-axis grinding (flank milling) and plunge milling have not been included.

The survey has been focused on the following categories: 1) *tool path interpolators*, 2) *adaptable geometric patterns* and 3) *methods for tool posture and gouging avoidance*. We believe that the above procedures are the most important part for efficient design of the numerical methods for five-axis machining.

Tool path interpolators

In the CNC machines, the tool motion is controlled by a sequence of reference points that are fed to the servo control system. The NC controllers employ linear interpolation techniques [66] or a circular interpolation which may result in discontinuities of the velocity at the junctions of the segments. They may also result in high accelerations, subsequent surface inaccuracies and long machining time required to eliminate them. Furthermore, the modern high speed machining requires feedrates up to 40 m/min with accelerations up to 2 g. At such high speeds, small discontinuities in the reference tool path can result in undesirable high frequency harmonics in the reference trajectory, which may end up exciting the natural modes of the mechanical structure and the servo control system.

Although the NC program cannot change the way the controller moves the machine parts, the cutter location points and the rotation angles required to cut the prescribed curve can be changed in such a way that these errors are minimized or at least decreased. The task of generating such a sequence of points is called interpolation. Early interpolation schemes solved the problem of discontinuities by smoothing the tool path at the corners [17] and using low pass filters [126]. However, the problem is due to the difference between the chord and the arc lengths. That is why, generating the tool positions by incrementing the chord length leads to the feedrate instabilities. Therefore, if the fit curve is parameterized with respect to the arc length, these type of the inaccuracies will be eliminated. Therefore, many modern interpolation schemes are focused on finding a suitable polynomial interpolation (such as the B-splines) parameterized by means of the arc length.

Unfortunately, such parameterization is not analytically possible for general spline curves. Therefore, a number of approximate solutions were proposed. For instance, Wang and Yang [122] generate the trajectory by means of cubic and quintic splines using the chord length and a *nearly arc length parameterization*, Zhang and Greenway [133] implemented a similar B-spline based interpolation. Coordinate transformations between the workpiece and machine coordinate systems for five-axis milling were incorporated in the interpolator by Lo [77, 79] and Bohez et al. [14]. An extra jerk continuity condition has been included into the solution in [121].

Furthermore, the limitations of the machine tool drivers may cause failure in maintaining the commanded feedrate which in turn may lead to the tool chatter or breakage. Therefore, Weck et al. [125] have implemented cubic spline interpolation where adaptation of the feedrate was based on the physical limitations of the drives. The smooth transitions were obtained using fourth order acceleration profiles. Erkorkmaz and Altintas [30] presented a quintic spline trajectory generation algorithm that produced continuous position, velocity, and acceleration profiles. Smooth accelerations and decelerations were provided by imposing constrains on the first and second time derivatives of the feedrate.

In 1994, Farouki and Sakkalis [33] introduced the Pythagorean-Hodograph (PH) curves to solve the problem of feedrate control for three-axis machines. The curves provide a mathematically elegant solution to the above mentioned problems occurring in NC machining. In particular, the arc length was represented by a polynomial function of the curve parameter. In [32], a 2D Hermite interpolation combined with the PH was proposed and analyzed. The ideas were further developed in [34, 35]. It was shown that since the arc length of the PH curves can be represented by a polynomial function of the curve parameter, they can be successfully used for the interpolation. Consequently, a variety of planar PH curves matching given Hermite type boundary data were developed (see, for instance, [61, 88, 119]).

Müller et al. [90] presented an algorithm for simultaneous five-axis spline interpolation which merges the PH interpolation and the analytic solution of the inverse kinematics problem using the template equation method. The result is a time-dependent spline which represents the given tool path with a high accuracy. Langeron et al. [69] suggested a polynomial B-spline interpolation which took into account the kinematics of the five-axis machine. The B-spline interpolation of the tool path in the part coordinate system includes the accuracy requirements and describes a five-axis tool path in a format adapted to the communication between the CAM software and the NC unit. The CAM output is directly expressed through the B-spline curves. Lo [80] introduced spline interpolators for *isoparametric, iso-scallop* and *iso-planar* machining methods (see the forthcoming section). Šír et al. [118] presented biarc interpolation techniques based on spline curves composed of circular arcs and compared them with the PH curves.

Finally, a number of papers introduce interpolators designed for high speed milling. During the high speed machining the actual average feedrate could be significantly lower than the programmed feedrate due to the physical restrictions of the machine tool and the block processing time of the CNC controller. In many cases the machine tool hardly reaches the maximum feedrates offered by the manufacturer. This happens when the block processing time is longer than the block execution time and the machine reaches the end point of the segment before information required for the next movement is available. In this case modern CNCs automatically reduce the programmed feedrate which results in a lower real feedrate and, consequently, a longer machining time. This relatively new issue has been discussed in [55, 64, 87, 109].

Adaptable geometric patterns

This section surveys research aimed to construct geometric patterns adaptable to a criteria which represents a certain estimate of the quality of the tool path such as the kinematics error, scallop heights, undercuts, overcuts, etc. It also includes methods for complex pocket milling since they often require special geometric patterns. Finally, the construction of the geometric patterns might or might not take into account the actual machine kinematics. In many cases, patterns employed for three-axis machining are also applicable, with certain modifications, to the five-axis machining.

The simplest tool path planning algorithms employ structured zigzag or spiral patterns due to their simplicity and the ease of computation (see [25, 108]). The zigzag and spiral motions employ uniform steps along a coordinate which parametrizes the desired curve extracted from the part surface.

The early adaptable methods replace the uniform spacing in favor of distributing the cutter location points by analogy with interpolation characterized by a variable step. First, the trajectories were assumed to be linear. Next, the desired curve was approximated using a certain technique, for example, employing arcs. Next, the forward step was selected by considering the deviation between the approximation and the straight line (see, for instance, [73]). The choice of the forward step can be performed by bisection or another inexpensive method. Some of the recent developments of these ideas are presented in [22, 73].

As mentioned before the most popular geometric solutions are the *zigzag* and the *spiral isoparametric* patterns constructed for single-patch or multi-patch parametric surfaces $S(u, v)$. In this context, the term *isoparametric* means that the zigzag tool path is generated in the parametric space $u - v$ along one of the coordinates, say, u. The v coordinate is then used to generate the forward steps.

Another approach is the *contour based* or *iso-planar* machining. In this case the cutter path follows intersection curves of the parametric surface and a series of vertical planes. One of the first papers reporting such techniques is [19], see also [104].

The both methods calculate the maximum allowable distance between the consecutive tracks using a scallop height limitation. However, if the maximum allowable distance is calculated globally, that is, the minimal allowable distance is taken from all the maximum allowable (pointwise) distances, then the method does not produce a constant scallop height. As the result, the machining efficiency is limited. Methods to maintain the constant scallop height called the *iso-scallop* machining methods were first proposed in [76, 112].

Lo [78] developed these approach and adapted it to five-axis machining. His algorithm starts with an initial curve in the parametric domain and calculates offset curves so that the scallop height remains approximately constant. The algorithm is designed for flat-end cuter and includes adaptive inclination which maximizes the machining strip. The algorithm also includes *a local gouging avoidance.* The local gouging refers to the removal of an excess material in the vicinity of the cutter contact point (CC point) due to the mismatch in curvatures between the tool as it is carried along the tool path and the desired surface (see the forthcoming section for details). Rao and Sarma [103] introduced similar *local gouging* avoidance algorithms applicable to surfaces characterized by low curvature and cut by the flat-end cutter. Finally, Lo [80] presents an iso-scallop tool path for *ball nose* cutters.

Evaluation of the machining strip versus the inclination and the direction could lead to complicated tool path topologies. For each tool position on the surface there exists at least one direction which maximizes the machining strip. The corresponding set of vectors mapped onto the parametric space (u, v) constitutes a 2D vector field which could be further analyzed. A continuous tool path which visits every point and follows the optimal direction at every point constitutes the optimal tool path which will maximize the machining strip globally. However, such a path can rarely be found in practice due to the complexity of the resulting vector field. An algorithm to find a suboptimal solution of this problem is presented in [20]. The authors introduce an "initial" tool path which has the largest average machining strip. Next, the entire tool path is constructed by propagating the initial path inside the region until it substantially deviates from the streamlines.

A few papers explore other "iso" methods such as the iso-distance and the iso-curvature methods (see, for instance, [47]). In [63] an additional tool path segments are appended to the basic tool path in order to achieve constant cutting forces and to avoid chatter vibrations in the entire machining area. Furthermore, the necessity to create geometric patterns suitable for the so-called pocket milling leaded to a series of methods designed for parts with one or more complex shaped "islands" inside. The methods are also needed to machined the so-called trimmed surfaces when the boundaries of the surface are defined by intersections with other surfaces.

In 1998, Choi and Jerard [21] introduced a term *regional milling* referring to situations when the machining operation, occurs in a region specified by boundary curves. According to [96] the regional milling can be performed using the same types of tool path topologies, namely, the *contour-based offset type* and the *direction-parallel type*. The *direction-parallel type* has been analyzed in [45, 46, 49], whereas the *direction-parallel type* in [21, 48, 51, 94].

One of the most important problems in the contour-based machining is linking the contours in such a way that the number of tool retractions is minimized. Held et al. [51] presents an algorithm designed for this type of machining based on the proximity maps and the Voronoi diagrams. The author suggests a linking procedure requiring a spanning tree of the planar graph of the monotonic pouches. Park and Chung [95] propose a contour linking algorithm accommodating minimization of slotting, tool-retractions and drilling holes.

Park et al. [96] presented a tool path linking algorithm, which guarantees a "zero" number of tool-retractions. The algorithm employs the concept of tool path element net providing information on the parent/child relationships. Jeong and Kim [57] present an algorithm designed to offset the boundary curves in the complex shaped region using the Voronoi diagram. Each curve segment is offset within the corresponding Voronoi polygon to avoid the degeneracy problem. Jeong and Kim [58] introduce a distance map algorithm which effectively finds the characteristic points and self intersection points of the offset curve segments and as the result eliminates such topological problems as loops, ridges and cusps. A forward locus tracing method is introduced in [68]. The algorithm searches for all intervals split by intersections of the planar curves and maps the 2D transversal intersections onto a 1D interval.

Suppose that the part is partitioned into a grid of cells each of them being a curvilinear triangle or a curvilinear rectangle. The optimization can then be considered as constructing a path which visits each cell, does not have intersections, requires minimal number of tool retractions and satisfies some error related criteria. Pocket machining using staircase or window frame patterns were proposed in [97]. Hansen and Arbab [46] developed a scan line algorithm for generating NC tool paths for arbitrarily shaped flat bottom pockets with islands. Flat pocket machining based on grids was suggested by Bao and Yim [11].

Treating the tool path generation as a navigation problem on grids leaded to approaches exploiting the shortest path optimization and related techniques. Suh and Shin [110] developed a neural network model to obtain the tool path in rough pocket machining as a solution to the traveling salesman problem. A good mathematical analysis of such strategies is given in [9]. The problem is formulated as follows: given a region in the parametric plane and the shape of a cutter find a shortest tour/path for the cutter such that every point within the region is covered by the cutter at some position along the tour (tool path). Additionally the cutter could be constrained to stay within a certain region. Narayanaswami and Choi [91] present a grid-based 3D navigation approach for generating NC tool path data for both linear interpolation and a combination of linear and circular interpolation for three-axis milling. The approach can be extended to the five-axis case.

The space-filling curves (SFC), having been applied in computer graphics, image processing, information systems, can be also seen as a suitable navigation pattern for generation of five-axis paths. The first application of the SFC to NC tool path generation was reported in [23, 44]. Griffiths [44] proposed the use of the Hilbert's curve as a tool path, while Cox et al. [23] used various forms of space-filling curves such as Moore's curve. Fractal based techniques were suggested by Chen et al. [18].

However, neither SFCs nor fractals have never been very popular in the five-axis machining community due to a large number of sharp turns produced by the conventional SFCs. A concept of an adaptive space-filling curve for tool path planning for five-axis NC machining was proposed in [8]. The space-filling curves, adapted to the local optimal cutting direction, produce shorter tool paths. Besides, the tool path correction stage suggested in [8] makes it possible to eliminate the effect of sharp angular turns which characterize the standard SFC patterns. These techniques will be presented in Chap. 4 of this book.

Finally, the entire tool path can be considered in the framework of the grid generation technologies. The concept was first introduced by Makhanov [82] and developed in [15, 83]. The grid generation techniques are surprisingly well-adapted to tool path optimizations. As a matter of fact, the concept of a grid refinement contains almost all the main ingredients for tool path planning, such as grid adaptation to the regions of large milling errors, possibility to easily construct curvilinear versions of the conventional zigzag and spiral patterns and adaptation to constraints related to the tool diameter and the scallop height. Moreover, in contrast to the standard techniques characterized by a local error estimate, grid generation deals with a global spatial error and consequently adapts all the CL points simultaneously. These ideas were developed further in [84], specifically for five-axis machining whereas Bieterman and Sandstrom [12] suggested a similar approach, independently. Finally, Sun et al. [111] presented a spiral version of the grid generation algorithm applied to tool path generation. The advanced grid generation techniques for five-axis machining will be considered in Chap. 5 of this book.

Tool posture and gouging avoidance

This section deals with techniques providing an optimal position and orientation of the tool in a particular neighborhood of the machined surface. The emphasis is on the orientations prescribed independently with regard to a certain criteria such as the local gouging avoidance, machining strip, scallop height, avoiding global gouging constraints, etc. The most important application of these techniques is cutting the part surface by the flat-end or fillet mill.

In 1987, Marciniak [86] showed that in five-axis machining the maximum width of machined strip on the surface could be obtained if the tool moved on the surface approximately along the minimum curvature line. The maximum width of the strip depends on the difference of the surface main curvatures at the contact point. Some of the early research papers exploiting this idea are [37, 67, 74]. Furthermore, Kruth and Klewais [67] introduced an optimal milling direction parallel to the principal direction of the surface with the minimum curvature.

Gani et al. [37] notice that "One of the critical problems in five-axis milling is the positioning of the cutter in relation to the surfaces in order to machine without having overcut (gouging) or undercut. Because of this problem, ball-end cutters are preferred. Undercutting does not cause a big problem when using ball-end cutters. The calculation of the NC tool path for ball-end cutters is mainly a problem of surface offset. An important drawback of ball-end cutters is the varying cutting speed along the tool radius. The maximal cutting speed is reached on the tool diameter, and at the tool tip it is zero. This leads to cutting edge chipping as well as poor surface roughness".

Recall that as long as the five-axis machines are considered, the tool has five degrees of freedom relative to the surface. The three spatial degrees are used to locate the tool at the cutter location points point. The extra two rotational degrees are used to establish the orientation of the tool represented by the *inclination angle* and the *tilt angle* (see, for instance, [103]) or the *tilt angle* and the *yaw angle* [60]. The angles are evaluated in a local coordinate system usually defined by the feed direction, the surface normal and the corresponding cross product vector. In the case of the flat-end mill the boundary of the base of the tool, which is the part of the tool cylinder is called the *cutting circle* of the tool. The *effective cutting shape* (also referred to as the *tool swept section*) is defined as a projection of the base of the tool onto the plane normal to the feed direction. Actually, in the case of flat-end mill the projected bottom edge becomes an ellipse called the *effective cutting ellipse*.

The parameters of the ellipse depend on the tool orientation. The *local gouging* (or the *curvature interference*) is usually defined as the excess material removal in the vicinity of the cutter contact point due to the mismatch in curvatures between the tool cutting edge and the desired surface. Detecting and avoidance of the local gouging includes comparing the *curvature of the effective cutting shape* (also referred to as *effective cutting curvature*) and the normal curvature of the surface evaluated in the same plane where the effective cutting shape is defined.

If the effective cutting curvature is greater than the normal curvature of the surface then the local gouging will not occur. The mathematical description of the effective cutting shape for a flat-end cutter is given in [129] and for a fillet end mill in [71]. The gouging is then avoided by determining the smallest inclination angle that ensures the largest material removal, that is, the largest machining strip. Of course, gouging is still possible because the curvatures are compared only in one section. In order to eliminate this source of errors, Lee and Ji [72] suggested to compare the curvature of the effective cutting shape evaluated in two planes: along the tool path and normal to the tool path. These effective cutting curvatures are compared to the normal curvatures of the surface in the respective planes and the inclination angle is computed as the maximum from the two minimal inclinations. Unfortunately, the method is not applicable to the non-convex surfaces when the radius of the curvature of the part surface is negative in the both directions but the maximum principle curvature is positive. In these cases, the method produces a zero inclination. This "bug" often leads to local gouging. Lo [78] solves this problem by continuously checking for gouging in all directions. Some improvements and modifications of these techniques are given in [8].

Li and Chen [75] write "Not only the parameters of the part of cutter body that pierces into the stock, but also the parameters of the area on the designed surface that may have relations to the cutter is yet to be studied. But the cutter location point, just as its name, is only the common point both on the cutter and the designed surface, any methods only based on the geometric properties of it will not obtain the best cutter positions." In other words, the accuracy of the above single point gouging model may be insufficient. In this case multipoint strategies [123, 124, 130] could be applied to further enhance the accuracy of the tool positioning. Furthermore, Rao and Sarma [103] present a closed form, coordinate free method for the detection and elimination of local gouging, at a CC point, in five-axis machining of sculptured surfaces using flat-end tools. The method is based on finding the curvatures of the tool swept surface at CC points along the tool path. Local gouging can then be detected and eliminated by sampling a finite set of points on the tool path, while comparing curvatures of the tool swept surface and the designed surface. Pottmann et al. [100] proposed a local millability criterion that guarantees global millability (i.e., rear-gouge and collision free milling) for three-axis machining using ball-end tools. The local millability criterion is based on curvature matching, using Dupin indicatrices in the tangent plane at the CC point, between the designed surface and the tool swept surface. A five-axis version of these method is presented in [131].

The above curvature matching methods require iterative gouge checking and correction strategies. To eliminate the need for iterative gouge checking and correction algorithms, some five-axis tool positioning strategies attempt to match the tool's cutting geometry to the surface geometry such as the principal axis method [104] developed for three-axis ball nose cutter and the five-axis arc intersection methods [42] based on the fact that the widest machined strip width is cut when the tool is tilted along the feed direction. Then for a given feed direction, a tool position is computed for each CC point along the tool path. The tool vector is restricted to lie in the tilting plane. The tilt angles are measured around the cross vector. The idea is to find the minimum tilt angle of the tool axis about the cross vector at which the tool contacts another point on the surface and maintain its contact with the CC point without gouging the surface.

Gray et al. [43] propose a modification of the five-axis arc intersection method for the so-called $3^1/_2{}^1/_2$-axis machining. This type of machining is characterized by three linear axes and two temporary locked rotary axes. The rotary axis are locked during the entire cut (resulting in a fixed tool orientation) or during a certain fraction of the cut. The rotary axis, represented by a high precision indexing device, constitute an interesting inexpensive alternative to the five-axis machining.

It should be noted that the above methods are based on the properties of a single or several contact points and therefore the errors still are unavoidable during, for instance, the wide strip precision machining. Besides, there is always a possibility of the so-called rear gouging when the back side of the tool gouges the surface with an attempt to obtain a wider machining strip. When the gouge is detected the tool must be inclined further and checked for gouging again until it clears the part. This secondary check and gouge elimination can be performed using the rolling ball method suggested in [41]. The basic idea is to roll a varying radius ball along the tool path and position the tool inside the ball.

An interesting approach has been developed by Li and Chen [75]. An envelop surface created by the cutter movement is discretized into infinite characteristic curves. Each of these curves will exactly copy themselves on to the stock. Then an analysis of the characteristic curves is performed to solve the problem of cutter positioning. The authors use the concept of the instantaneous cutter position error employing the virtual cutting edge of the tool. The effective bandwidth of cutting strip is calculated and used in the optimization algorithm.

Finally, regardless of the tool orientation there always exists the possibility of a global interference of the workpiece with the tool holder, fixture or other parts of the machine. There also exists a possibility that given any orientation, the tool still flank-mills an unwanted part of the surface. It means that this portion of the surface is not accessible. Elber and Cohen [28] write "The problem of accessibly, or the ability to verify and possibly correct gouging into the machined surface or even into other surfaces, is apparently the most fundamental hindering factor in the broad use of five-axis machining".

Solid modeling systems offer the possibility of doing both simulation and verification of tool paths off-line. However, the solid modeling approach is computationally expensive. The cost of simulation using the so-called constructive solid geometry is proportional to the fourth power of the number of the tool movements $O(N^4)$ [16]. On the other hand, the solid model can detect both the local and the global interference, including collisions with the clamping device and the machine parts. A typical program for surface machining could contain more than 10,000 tool movements, therefore, current solid modeling research focuses on efficient and fast algorithms to compute the swept volume of the tool and perform Boolean operations to subtract the intersection from the stock. Bohez et al. [16] presents a short introduction to solid modeling schemes such as the extended Z-buffer algorithm [52], line graphic simulation approaches [59] and others.

The partition into elements and the corresponding data structures are the most important components of these procedures. The Z-buffer structure [7, 113], ray representation [56], Octree method [93, 106], K-D trees [50], BSP-trees [92], Brep-indices [65, 117], tetrahedral meshes [93] and regular grids [38] are examples of such spatial decomposition techniques. Each solid modeling algorithm has advantages and disadvantages in terms of accuracy, robustness, data structure and computation time. However, it seems that the simplicity of the data structure required for the Z-buffer scheme and the possibility to generate and update the part model very fast made many commercial CAM program to use the Z-buffer algorithm or its ramifications for the NC code verification and optimization [107].

The recent research papers include many improvements of the Z-buffer techniques such as the enhanced Z-buffer model [107], the stencil buffer [16], the adaptive depth buffer [105], the undo facilities for the Z-buffer scheme [13], etc. However, the above methods are not designed specifically for global interference detection. As a matter of fact, the solid model visualizes a general cut which may or may not include the global interferences. It may take hours of simulation and possibly an operator to visually detect possible collisions.

Therefore, methods based on the closed form mathematical solutions or their approximation are still valuable. The problem of the global gouging is treated mathematically using the concept of *accessibility*. The *accessibility of a point in a given direction* is defined as follows: a point belonging to a geometric entity is *accessible* in a given direction if a ray can be drawn from it in the given direction without intersecting with interior of the geometric entity.

The problem of accessibility in three-axis machining can be solved by a method of hidden surface removal of the same scene from a direction collinear with the tool axis [26, 53]. The fact that the tool has a finite thickness can be compensated for, by offsetting all the check surfaces by the radius of the tool.

The use of Z-buffer based hidden surface removal techniques to verify and correct the three-axis tool path, is a common practice in many contemporary computer-aided manufacturing schemes. For three-axis machining under certain conditions the absence of the local gouging implies the complete absence of collisions [99].

For five-axis machining it has been also shown that if all axis positions pass through a fixed point and if all points of the workpiece surface can be seen from this point then the local millability implies global millability [99]. Wallner and Pottmann [120] presented a global millability theorem for general workpieces. They analyzed several possible configuration manifolds of tool positions relative to a workpiece under different aspects; the degree of freedom of the motion of the tool, the correspondence between the contact point and the tool position, and the presence or absence of unwanted collisions between tool and workpiece.

Takeuchi et al. [114] proposed a method for computing the collision-free CL data using a trial and error approach. Morishige et al. [89] used the so-called C-space techniques to generate a smooth, continuously varying tool path. The C-space is a general concept of robotics where the configuration of a mechanism is specified by a sequence of values. A rigid body, for example, can be located in space by specifying six parameters related to all six of its degrees of freedom. The configuration space (C-space) of a mechanism is the space of these parameters, and a point in the C-space specifies a particular configuration. Obstacles can be mapped to the C-space as well, and the required collision-free access can theoretically be inferred by navigating the point in the C-space around the obstacles. Unfortunately, though intuitive and intellectually appealing, the C-space approach could lead to computationally intractable tasks.

The problem of accessibility can be approximated by a simpler requirement called visibility. A point on an object is visible from a point at infinity if there exists a straight line segment connecting the two points which does not intersect with the object. Visibility is a useful precursor for the accessibility computation because, for a certain class of tools and probes, visibility is a necessary condition for accessibility. Seminal theoretical results in the area were obtained by Elber and Cohen [27], Elber and Zussman [29], Woo [127]. Later research was focused on generating the so-called product visibility cones (see [62, 116]).

Lauwers et al. [70] describe a multi-axis tool path generation software where the tool orientation is optimized to avoid machine collisions and at the same time to maximize the material removal rate along the tool track. To perform efficient collision avoidance, the tool path generation module, the post processing and machine simulation has been integrated into one system. Xu et al. [128] combine the machine limits, collisions, and gouging to generate feasible gouge free tool orientations. Gian et al. [39] developed open regions and vector fields techniques to find rapidly the cutter paths and tool orientations for parts with cavity areas. In 2003, Balasubramaniam et al. [10] developed methods for five-axis tool positioning that account for accessibility of the tool using visibility maps of the triangulated data. Using this visibility data for finish machining the authors show how it can be used to generate globally collision-free five-axis finishing tool paths while considering machine limits, tool tilt, cusp height limits and the tool pitch limits.

Young et al. [132] presented a new parametric method with an approximate constant cutting depth for the rough machining of an impeller. The initial tool spindle axis is considered as the initial orientation to determine the cutting tool posture for which the variation of rotational axes of the five-axis machine tool will be reduced. Hsueh et al. [54] propose to prevent the collisions using the two stages: the first stage is to obtain the tilting and collision-free angle range in the plane that is normal to the tool path obtained. Next, a checking cone generated from this collision-free tool axis range is used for the second collision check. The collision region is formed by the intersection of the neighboring surfaces.

Analyzing a proper sculptured surface orientation on the worktable of multi-axis CNC machine, Radzevich and Goodman [102] proposed the so-called spherical indicatrix of the sculptured surface machinability. This characteristic curve indicates whether the sculptured surface is machinable under a known scenario. The theory is developed in connection with a sculptured surface orientation on the worktable of a multi-axis CNC machine.

Radzevich [101] presents an approach that enables us to detect regions of a sculptured surface which are not accessible for a cutting tool of a given design, Furthermore, if any not-machinable regions exist, the developed approach enables us to subdivide the sculptured surface into the cutter-accessible and the cutter-not-accessible regions.

We have presented a survey on three selected topics in five-axis machining, namely, tool path interpolators, adaptable geometric patterns and methods for tool posture and gouging avoidance. We believe that the above procedures are the most important for constructing efficient numerical methods for five-axis machining

It should be noted that complete software systems designed for tool path generation and verification are now very important. The commercial CAD/CAM systems such as Unigraphics, PowerMill, Catia, MasterCam, each of them to a certain degree, include five-axis capabilities. In Chap. 2 we will demonstrate how the solid modeling features of Unigraphics can be used for the tool path simulation and verification.

Finally, we apologize to those authors whose works have not been cited. The exclusion of any such papers is due to our not being aware of their work.

References

[1] The NURBS++ package, http://libnurbs.sourceforge.net/index.shtml.
[2] NURBS Toolbox, http://www.aria.uklinux.net/nurbs.php3.
[3] NIST/IGES, http://www.nist.gov/iges/.
[4] .SLC file format, http://www-rp.me.vt.edu/bohn/rp/SLC.html.
[5] CAD/CAM Software and File Format Compatiblity Matrix, http://www.cs.cmu.edu/People/unsal/research/rapid/cadcam.html.
[6] CAD File Formats, http://www.actify.com/v2/products/Importers/formats.htm.
[7] Anderson, R. O. 1978. Detecting and eliminating collisions in NC machining. *Computer-Aided Design*, 10(4):231–237.
[8] Anotaipaiboon, W. and Makhanov, S. S. 2005. Tool path generation for five-axis NC machining using adaptive space-filling curves. *International Journal of Production Research*, 43(8):1643–1665.
[9] Arkin, E. M., Fekete, S. P., and Mitchell, J. S. B. 2000. Approximation algorithms for lawn mowing and milling. *Computational Geometry: Theory and Applications*, 17(1-2):25–50.
[10] Balasubramaniam, M., Sarma, S. E., and Marciniak, K. 2003. Collision-free finishing toolpaths from visibility data. *Computer-Aided Design*, 35(4):359–374.
[11] Bao, H. P. and Yim, H. 1992. Tool path determination for end milling of non-convex shaped polygons. *NAMRI Transactions*, pages 151–158.
[12] Bieterman, M. B. and Sandstrom, D. R. 2003. A curvilinear tool-path method for pocket machining. *Journal of Materials Processing Technology*, 125(4):709–715.
[13] Blasquez, I. and Poiraudeau, J.-F. 2004. Undo facilities for the extended z-buffer in NC machining simulation. *Computers in Industry*, 53(2):193–204.
[14] Bohez, E. L. J., Makhanov, S. S., and Sonthipermpoon, K. 2000a. Adaptive non-linear grid tool path optimization for 5-axis machining. *International Journal of Production Research*, 38(17):4329–4343.
[15] Bohez, E. L. J., Makhanov, S. S., and Sonthipermpoon, K. 2000b. Adaptive nonlinear tool path optimization for 5-axis machining. *International Journal of Production Research*, 38(17):4329–4343.
[16] Bohez, E. L. J., Minh, N. T. H., Kiatsrithanakorn, B., Natasukon, P., Ruei-Yun, H., and Son, L. T. 2003. The stencil buffer sweep plane algorithm for 5-axis CNC tool path verification. *Computer-Aided Design*, 35(12):1129–1142.
[17] Butler, J., Haack, B., and Tomizuka, M. 1988. Reference input generation for high speed coordinated motion of a two axis system. In *Symposium on Robotics, Winter Annual Meeting of the American Society of Mechanical Engineers*, pages 457–470.

[18] Chen, C.-C. A., Juang, Y.-S., and Lin, W.-Z. 2002. Generation of fractal toolpaths for irregular shapes of surface finishing areas. *Journal of Materials Processing Technology*, 127(2):146–150.
[19] Chen, Y. D., Ni, J., and Wu, S. M. 1993. Real-time CNC tool path generation for machining IGES surfaces. *ASME Journal of Engineering for Industry*, 115(4):480–486.
[20] Chiou, C.-J. and Lee, Y.-S. 2002. A machining potential field approach to tool path generation for multi-axis sculptured surface machining. *Computer-Aided Design*, 34(5):357–371.
[21] Choi, B. K. and Jerard, R. B. 1998. *Computer Aided Machining - the z-Map Way: Sculptured Surface Machining - Theory and Applications.* Kluwer Academic Publishers, Dordrecht, The Netherlands.
[22] Choi, Y.-K. and Banerjee, A. Tool path generation and tolerance analysis for free-form surfaces. *International Journal of Machine Tools and Manufacture.* in press.
[23] Cox, J. J., Takezaki, Y., Ferguson, H. R. P., Kohkonen, K. E., and Mulkay, E. L. 1994. Space-filling curves in tool-path applications. *Computer-Aided Design*, 26(3):215–224.
[24] De Boor, C. 2001. *A practical guide to splines.* Springer, New York, USA.
[25] Dragomatz, D. and Mann, S. 1997. A classified bibliography of literature on NC milling path generation. *Computer-Aided Design*, 29(3):239–247.
[26] Elber, G. and Cohen, E. 1990. Hidden curve removal for free form surfaces. In *SIGGRAPH '90: Proceedings of the 17th annual conference on Computer graphics and interactive techniques*, pages 95–104.
[27] Elber, G. and Cohen, E. 1995. Arbitrarily precise computation of gauss maps and visibility sets for freeform surfaces. In *SMA '95: Proceedings of the third ACM symposium on Solid modeling and applications*, pages 271–279.
[28] Elber, G. and Cohen, E. 1999. A unified approach to verification in 5-axis freeform milling environments. *Computer-Aided Design*, 31(13):795–804.
[29] Elber, G. and Zussman, E. 1998. Cone visibility decomposition of freeform surface. *Computer-Aided Design*, 30(4):315–320.
[30] Erkorkmaz, K. and Altintas, Y. 2001. High speed CNC system design. Part I: jerk limited trajectory generation and quintic spline interpolation. *International Journal of Machine Tools and Manufacture*, 41(9):1323–1345.
[31] Farin, G. E. 1999. *NURBS: From Projective Geometry to Practical Use.* A. K. Peters, Ltd., Natick, MA, USA.
[32] Farouki, R. T. and Neff, C. A. 1995. Hermite interpolation by Pythagorean hodograph quintics. *Mathematics of Computation*, 64(212):1589–1609.

[33] Farouki, R. T. and Sakkalis, T. 1994. Pythagorean-hodograph space curves. *Advances in Computational Mathematics*, 2(1):41–66.
[34] Farouki, R. T., Tsai, Y.-F., and Wilson, C. S. 2000. Physical constraints on feedrates and feed accelerations along curved tool paths. *Computer Aided Geometric Design*, 17(4):337–359.
[35] Farouki, R. T., Tsai, Y.-F., and Yuan, G.-F. 1999. Contour machining of free-form surfaces with real-time PH curve CNC interpolators. *Computer Aided Geometric Design*, 16(1):61–76.
[36] Flutter, A. and Todd, J. 2001. A machining strategy for toolmaking. *Computer-Aided Design*, 33(13):1009–1022.
[37] Gani, E. A., Kruth, J. P., Vanherck, P., and Lauwers, B. 1997. A geometrical model of the cut in five-axis milling accounting for the influence of tool orientation. *International Journal of Advanced Manufacturing Technology*, 13(10):677–684.
[38] Garcia-Alonso, A., Serrano, N., and Flaquer, J. 1994. Solving the collision detection problem. *IEEE Computer Graphics and Applications*, 14(3):36–43.
[39] Gian, R., Lin, T., and Lin, A. C. 2003. Planning of tool orientation for five-axis cavity machining. *International Journal of Advanced Manufacturing Technology*, 22(1-2):150–160.
[40] Goldstein, B. L., Kemmerer, S. J., and Parks, C. H. 1998. A brief history of early product data exchange standards - NISTIR 6221. Technical report, National Institute of Standards and Technology.
[41] Gray, P. J., Bedi, S., and Ismail, F. 2003. Rolling ball method for 5-axis surface machining. *Computer-Aided Design*, 35(4):347–357.
[42] Gray, P. J., Bedi, S., and Ismail, F. 2005. Arc-intersect method for 5-axis tool positioning. *Computer-Aided Design*, 37(7):663–674.
[43] Gray, P. J., Ismail, F., and Bedi, S. 2007. Arc-intersect method for $3\frac{11}{22}$-axis tool paths on a 5-axis machine. *International Journal of Machine Tools and Manufacture*, 47(1):182–190.
[44] Griffiths, J. G. 1994. Toolpath based on Hilbert's curve. *Computer-Aided Design*, 26(11):839–844.
[45] Guyder, M. K. 1990. Automating the optimization of 2 1/2 axis milling. *Computers in Industry*, 15(3):163–168.
[46] Hansen, A. and Arbab, F. 1992. An algorithm for generating NC tool paths for arbitrarily shaped pockets with islands. *ACM Trans. Graph.*, 11(2):152–182.
[47] Hatna, A. and Grieve, B. 2000. Cartesian machining versus parametric machining: a comparative study. *International Journal of Production Research*, 38(13):3043–3065.
[48] Held, M. 1991a. A geometry-based investigation of the tool path generation for zigzag pocket machining. *The Visual Computer*, 7(5-6):296–308.
[49] Held, M. 1991b. *On the computational geometry of pocket machining.* Springer-Verlag New York, Inc., New York, NY, USA.

[50] Held, M., Klosowski, J., and Mitchell, J. S. B. 1995. Evaluation of collision detection methods for virtual reality fly-throughs. In *proceedings Seventh Canadian Conference on Computational Geometry*, pages 205–210.

[51] Held, M., Lukács, G., and Andor, L. 1994. Pocket machining based on contour-parallel tool paths generated by means of proximity maps. *Computer-Aided Design*, 26(3):189–203.

[52] Hook, T. V. 1986. Real-time shaded NC milling display. In *SIGGRAPH '86: Proceedings of the 13th annual conference on Computer graphics and interactive techniques*, pages 15–20.

[53] Hornung, C., Lellek, W., Rehwald, P., and Straßer, W. 1985. An area-oriented analytical visibility method for displaying parametrically defined tensor-product surfaces. *Computer Aided Geometric Design*, 2(1-3):197–205.

[54] Hsueh, Y.-W., Hsueh, M.-H., and Lien, H.-C. Automatic selection of cutter orientation for preventing the collision problem on a five-axis machining. *International Journal of Advanced Manufacturing Technology*. in press.

[55] Hu, J., Xiao, L., Wang, Y., and Wu, Z. 2006. An optimal feedrate model and solution algorithm for a high-speed machine of small line blocks with look-ahead. *International Journal Advanced Manufacturing Technology*, 28(9):930–935.

[56] Huang, Y. and Oliver, J. H. 1995. Integrated simulation, error assessment, and tool path correction for five-axis NC machining. *Journal of Manufacturing Systems*, 14(5):331334.

[57] Jeong, J. and Kim, K. 1999a. Generating tool paths for free-form pocket machining using z-buffer-based Voronoi diagrams. *International Journal of Advanced Manufacturing Technology*, 15(3):182–187.

[58] Jeong, J. and Kim, K. 1999b. Generation of tool paths for machining free-form pockets with islands using distance maps. *International Journal of Advanced Manufacturing Technology*, 15(5):311–316.

[59] Jerard, R. B. and Drysdale, R. L. 1989. Methods for geometric modeling, simulation and spacial verification of NC machining programs. In Wozny, M. J., Turner, J. U., and Pegna, J., editors, *Product modeling for computer-aided design and manufacturing*. Elsevier/North-Holland, New York, USA.

[60] Jun, C.-S., Cha, K., and Lee, Y.-S. 2003. Optimizing tool orientations for 5-axis machining by configuration-space search method. *Computer-Aided Design*, 35(6):549–566.

[61] Jüttler, B. 2001. Hermite interpolation by Pythagorean hodograph curves of degree seven. *Mathematics of Computation*, 70(235):1089–1111.

[62] Kang, J.-K. and Suh, S.-H. 1997. Machinability and set-up orientation for five-axis numerically controlled machining of free surfaces. *International Journal of Advanced Manufacturing Technology*, 13(5):311–325.

[63] Kim, H.-C., Lee, S.-G., and Yang, M.-Y. An optimized contour parallel tool path for 2D milling with flat endmill. *International Journal of Advanced Manufacturing Technology*. in press.
[64] Ko, T. J., Kim, H. S., and Park, S. H. 2005. Machineability in NURBS interpolator considering constant material removal rate. *International Journal of Machine Tools and Manufacture*, 45(6):665–671.
[65] Kondo, M. 1994. Decomposition of complex geometry for a manufacturing application. *Computer-Aided Design*, 26(3):244–252.
[66] Koren, Y. 1976. Interpolator for a computer numerical control system. *IEEE Transactions on Computers*, 25(1):32–37.
[67] Kruth, J.-P. and Klewais, P. 1994. Optimization and dynamic adaptation of the cutter inclination during five-axis milling of sculptured surfaces. *Anals CIRP*, 43(1):443–448.
[68] Lai, Y.-L., Wu, J. S.-S., Hung, J.-P., and Chen, J.-H. 2006. A simple method for invalid loops removal of planar offset curves. *International Journal of Advanced Manufacturing Technology*, 27(11-12):1153–1162.
[69] Langeron, J. M., Duc, E., Lartigue, C., and Bourdet, P. 2004. A new format for 5-axis tool path computation, using Bspline curves. *Computer-Aided Design*, 36(12):1219–1229.
[70] Lauwers, B., Dejonghe, P., and Kruth, J. P. 2003. Optimal and collision free tool posture in five-axis machining through the tight integration of tool path generation and machine simulation. *Computer-Aided Design*, 35(5):421–432.
[71] Lee, Y.-S. 1997. Admissible tool orientation control of gouging avoidance for 5-axis complex surface machining. *Computer-Aided Design*, 29(7):507–521.
[72] Lee, Y.-S. and Ji, H. 1997. Surface interrogation and machining strip evaluation for 5-axis CNC die and mold machining. *International Journal of Production Research*, 35(1):225–252.
[73] Li, F., Wang, X. C., Ghosh, S. K., Kong, D. Z., Lai, T. Q., and Wu, X. T. 1995. Tool-path generation for machining sculptured surface. *Journal of Materials Processing Technology*, 48(1):811–816.
[74] Li, S. X. and Jerard, R. B. 1994. 5-axis machining of sculptured surfaces with a flat-end cutter. *Computer-Aided Design*, 26(3):165–178.
[75] Li, Z. and Chen, W. 2006. A global cutter positioning method for multi-axis machining of sculptured surfaces. *International Journal of Machine Tools and Manufacture*, 46(12-13):1428–1434.
[76] Lin, R.-S. and Koren, Y. 1996. Efficient tool-path planning for machining free-form surfaces. *ASME Journal of Engineering for Industry*, 118(1):20–28.
[77] Lo, C. C. 1997. Feedback interpolator for CNC machine tool. *ASME. Journal of Manufacturing Science and Engineering*, 119(4):587592.
[78] Lo, C. C. 1999a. Efficient cutter-path planning for five-axis surface machining with a flat-end cutter. *Computer-Aided Design*, 31(9):557–566.

[79] Lo, C. C. 1999b. Real-time generation and control of cutter path for 5-axis CNC machining. *International Journal of Machine Tools and Manufacture*, 39(3):471–488.

[80] Lo, C. C. 2000. CNC machine tool surface interpolator for ball-end milling of free-form surfaces. *International journal of machine tools and manufacture*, 40(3):307–326.

[81] Ma, W., But, W.-C., and He, P. 2004. NURBS-based adaptive slicing for efficient rapid prototyping. *Computer-Aided Design*, 36(13):1309–1325.

[82] Makhanov, S. S. 1999. An application of variational grid generation techniques to the tool-path optimization of industrial milling robots. *Journal of Materials Processing Technology*, 39(9):1524–1535.

[83] Makhanov, S. S., Batanov, D., Bohez, E., Sonthipaumpoon, K., Anotaipaiboon, W., and Tabucanon, M. 2002. On the tool-path optimization of a milling robot. *Computers & Industrial Engineering*, 43(3):455–472.

[84] Makhanov, S. S. and Ivanenko, S. A. 2003. Grid generation as applied to optimize cutting operations of the five-axis milling machine. *Applied Numerical Mathematics*, 46(3-4):331–351.

[85] Mani, K., Kulkarni, P., and Dutta, D. 1999. Region-based adaptive slicing. *Computer-Aided Design*, 31(5):317–333.

[86] Marciniak, K. 1987. Influence of surface shape in admissible tool positions in 5-axis face milling. *Computer-Aided Design*, 19(5):233–236.

[87] Monreal, M. and Rodríguez, C. A. 2003. Influence of tool path strategy on the cycle time of high-speed milling. *Computer-Aided Design*, 35(4):395–401.

[88] Moon, H. P., Farouki, R. T., and Choi, H. I. 2001. Construction and shape analysis of PH quintic hermite interpolants. *Computer Aided Geometric Design*, 18(2):93–115.

[89] Morishige, K., Takeuchi, Y., and Kase, K. 1999. Tool path generation using C-space for 5-axis control machining. *Journal of Manufacturing Science and Engineering*, 121(1):144–149.

[90] Müller, M., Erds, G., and Xirouchakis, P. C. 2004. High accuracy spline interpolation for 5-axis machining. *Computer-Aided Design*, 36(13):1379–1393.

[91] Narayanaswami, R. and Choi, Y. 2001. NC machining of freeform pockets with arbitrary wall geometry using a grid-based navigation approach. *International Journal of Advanced Manufacturing Technology*, 18(10):708–716.

[92] Naylor, B., Amanatides, J., and Thibault, W. 1990. Merging BSP trees yields polyhedral set operations. In *SIGGRAPH '90: Proceedings of the 17th annual conference on Computer graphics and interactive techniques*, pages 115–124.

[93] Noborio, H., Fukuda, S., and Arimoto, S. 1989. Fast interference check method using octree representation. *Advanced robotics*, 3(3):193–212.

[94] Park, S. C. and Choi, B. K. 2000. Tool-path planning for direction-parallel area milling. *Computer-Aided Design*, 32(1):17–25.

[95] Park, S. C. and Chung, Y. C. 2002. Offset tool-path linking for pocket machining. *Computer-Aided Design*, 34(4):299–308.

[96] Park, S. C., Chung, Y. C., and Choi, B. K. 2003. Contour-parallel offset machining without tool-retractions. *Computer-Aided Design*, 35(9):841–849.

[97] Persson, H. 1978. NC machining of arbitrarily shaped pockets. *Computer-Aided Design*, 10(3):169–174.

[98] Piegl, L. and Tiller, W. 1995. *The NURBS book*. Springer-Verlag, London, UK.

[99] Pottmann, H. and Ravani, B. 2000. Singularities of motions constrained by contacting surfaces. *Mechanism and Machine Theory*, 35(7):963–984.

[100] Pottmann, H., Wallner, J., Glaeser, G., and Ravani, B. 1999. Geometric criteria for gouge-free three-axis milling of sculptured surfaces. *ASME Journal of Mechanical Design*, 121(2):241–248.

[101] Radzevich, S. P. 2005. A cutting-tool-dependent approach for partitioning of sculptured surface. *Computer-Aided Design*, 37(7):767–778.

[102] Radzevich, S. P. and Goodman, E. D. 2002. Computation of optimal workpiece orientation for multi-axis NC machining of sculptured part surfaces. *ASME Journal of Mechanical Design*, 124(2):201–212.

[103] Rao, A. and Sarma, R. 2000. On local gouging in five-axis sculptured surface machining using flat-end tools. *Computer-Aided Design*, 32(7):409–420.

[104] Rao, N., Ismail, F., and Bedi, S. 1997. Tool path planning for five-axis machining using the principal axis method. *International Journal of Machine Tools and Manufacture*, 37(7):1025–1040.

[105] Roth, D., Ismail, F., and Bedi, S. 2003. Mechanistic modelling of the milling process using an adaptive depth buffer. *Computer-Aided Design*, 35(14):1287–1303.

[106] Roy, U. and Xu, Y. 1999. Computation of a geometric model of a machined part from its NC machining programs. *Computer-Aided Design*, 31(6):401–411.

[107] Sang-Kyu Lee, S.-L. K. 2002. Development of simulation system for machining process using enhanced Z map model. *Journal of materials processing technology*, 130-131:608–617.

[108] Sarma, R. 2000. An assessment of geometric methods in trajectory synthesis for shape-creating manufacturing operations. *Journal of Manufacturing Systems*, 19(1):59–72.

[109] Siller, H., Rodriguez, C. A., and Ahuett, H. 2006. Cycle time prediction in high-speed milling operations for sculptured surface finishing. *Journal of Materials Processing Technology*, 174(1-3):355–362.

[110] Suh, S. H. and Shin, Y. S. 1996. Neural network modeling for tool path planning of rough cut in complex pocket milling. *Journal of Manufacturing Systems*, 15(5):295–304.

[111] Sun, Y.-W., Guoa, D.-M., and Jia, Z.-Y. 2006. Spiral cutting operation strategy for machining of sculptured surfaces by conformal map approach. *Journal of Materials Processing Technology*, 180(1-3):74–82.

[112] Suresh, K. and Yang, D. C. H. 1994. Constant scallop-height machining of free-form surfaces. *ASME Journal of Engineering for Industry*, 116(2):253–259.

[113] Takata, S. 1989. A cutting simulation system for machinability evaluation using a workpiece model. *Anals CIRP*, 38(1):417–420.

[114] Takeuchi, Y., Shimizu, H., Idemura, T., Watanabe, T., and Ito, T. 1990. 5-axis control machining based on solid model. *Journal of the Japan Society for Precision Engineering*, 56(1):111–116.

[115] Tata, K., Fadel, G., Bagchi, A., and Aziz, N. 1998. Efficient slicing for layered manufacturing. *Rapid Prototyping Journal*, 4(4):151–167.

[116] Vafaeesefa, A. and ElMaraghy, H. A. 1998. Accessibility analysis in 5-axis machining of sculptured surfaces. In *Proceedings of the 1998 IEEE International Conference on Robotics & Automation*, pages 2464–2469.

[117] Vaněček, Jr., G. 1991. BRep-Index: a multidimensional space partitioning tree. In *SMA '91: Proceedings of the first ACM symposium on Solid modeling foundations and CAD/CAM applications*, pages 35–44.

[118] Šír, Z., Feichtinger, R., and Jüttler, B. 2006. Approximating curves and their offsets using biarcs and Pythagorean hodograph quintics. *Computer-Aided Design*, 38(6):608–618.

[119] Šír, Z. and Jüttler, B. 2005. Constructing acceleration continuous tool paths using Pythagorean hodograph curves. *Mechanism and Machine Theory*, 40(11):1258–1272.

[120] Wallner, J. and Pottmann, H. 2000. On the geometry of sculptured surface machining. In Laurent, P.-J., Sablonnière, P., and Schumaker, L. L., editors, *Curve and Surface Design*. Vanderbilt University Press, Nashville, USA.

[121] Wang, F.-C. and Wright, P. K. 1998. Open architecture controllers for machine tools, Part 2: A real time quintic spline interpolator. *Journal of manufacturing science and engineering*, 120(2):425–432.

[122] Wang, F.-C. and Yang, D. C. H. 1993. Nearly arc-length parameterized quintic-spline interpolation for precision machining. *Computer-Aided Design*, 25(5):281–288.

[123] Warkentin, A., Ismail, F., and Bedi, S. 1998. Intersection approach to multi-point machining of sculptured surfaces. *Computer Aided Geometric Design*, 15(6):567–584.

[124] Warkentin, A., Ismail, F., and Bedi, S. 2000. Multi-point tool positioning strategy for 5-axis machining of sculptured surfaces. *Computer Aided Geometric Design*, 17(1):83–100.

[125] Weck, M., Meylahn, A., and Hardebusch, C. 1999. Innovative algorithms for spline-based CNC controller, production engineering research and development in germany. *Annals of the German Academic Society for Production Engineering*, VI(1):83–86.

[126] Weck, M. and Ye, G. H. 1990. Sharp corner tracking using the IKF control strategy. *Annals CIRP*, 39(1):437–441.

[127] Woo, T. C. 1994. Visibility maps and spherical algorithms. *Computer-Aided Design*, 26(1):6–16.

[128] Xu, X. J., Bradley, C., Zhang, Y. F., Loh, H. T., and Wong, Y. S. 2002. Tool-path generation for five-axis machining of free-form surfaces based on accessibility analysis. *International Journal of Production Research*, 40(14):3253–3274.

[129] Y-S Lee, T.-C. C. 1996. Automatic cutter selection for five-axis sculptured surface machining. *International Journal of Production Research*, 34(4):977–998.

[130] Yoon, J.-H. 1997. Tool tip gouging avoidance and optimal tool positioning for 5-axis sculptured surface machining. *International Journal of Production Research*, 41(10):2125–2142.

[131] Yoon, J.-H., Pottmann, H., and Lee, Y.-S. 2003. Locally optimal cutting positions for 5-axis sculptured surface machining. *Computer-Aided Design*, 35(1):69–81.

[132] Young, H.-T., Chuang, L.-C., Gerschwiler, K., and Kamps, S. 2004. A five-axis rough machining approach for a centrifugal impeller. *International Journal of Advanced Manufacturing Technology*, 23(3-4):233–239.

[133] Zhang, Q. G. and Greenway, R. B. 1998. Development and implementation of a NURBS curve motion interpolator. *Robotics and Computer-Integrated Manufacturing*, 14(1):27–36.

2

Introduction to Five-Axis NC Machining

2.1 Five-Axis NC Machining Concepts

Numerical control or numerically controlled (NC) machines are automatically operated by commands received by their processing units. The NC machines were first developed soon after World War II and made it possible for a precise and efficient production of large quantities of the desired components in a reliable repetitive manner. Early NC machines were often fed with instructions which were punched onto paper tape or punch cards. In the 1960s, NC machines largely gave way to computer numerical control (CNC) machines which refers specifically to a computer *controller* that reads G-code instructions (see Sect. 2.2) and drives the machine tool.

Five-axis milling machines are NC machines which characterized by three translational and two rotary axes. A five-axis milling machine with two rotary axes on the table is depicted in Fig. 2.1. A block of a raw material called the workpiece is fixed to the *machine table* by means of *clamps* (*clumping device*) (see Fig. 2.1b). The material is then removed from the workpiece by a rotating *cutter* attached to the *spindle* through the *tool holder*. The process of material removal with the goal to produce a required industrial part is called *milling*, *machining* or *cutting*.

The NC machines are programmed by means of a special code called the *NC program* or the *part program* composed of commands represented by letters, numbers and special symbols. The part program consists of instructions to control the machine movements following a certain manufacturing technology and methodology. For example, the program tells the machine what is the required shape and size of the tool, at what speed and feedrate and at what orientation relative to the workpiece this tool should be used. The part program also prescribes a set of cutter location[1] (CL) points assigned to cut the desired part.

[1] A cutter location refers to the position at which an NC milling machine has been instructed to hold the milling cutter.

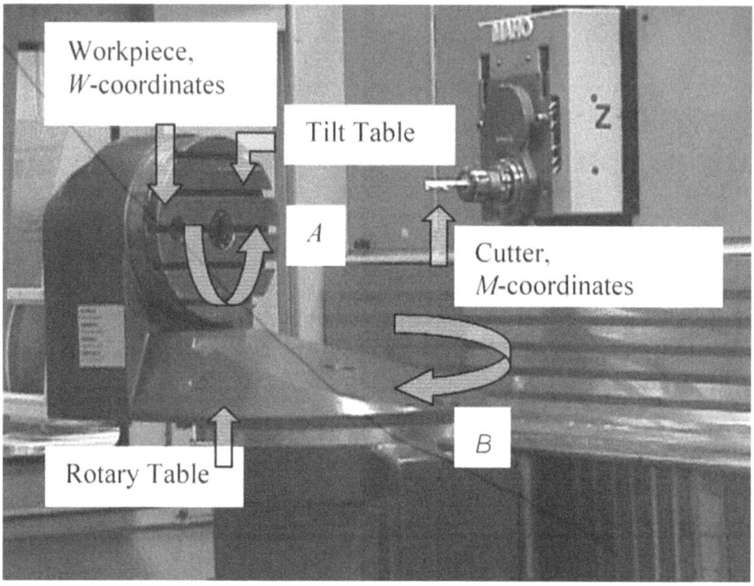

Fig. 2.1a. MAHO 600E milling machine

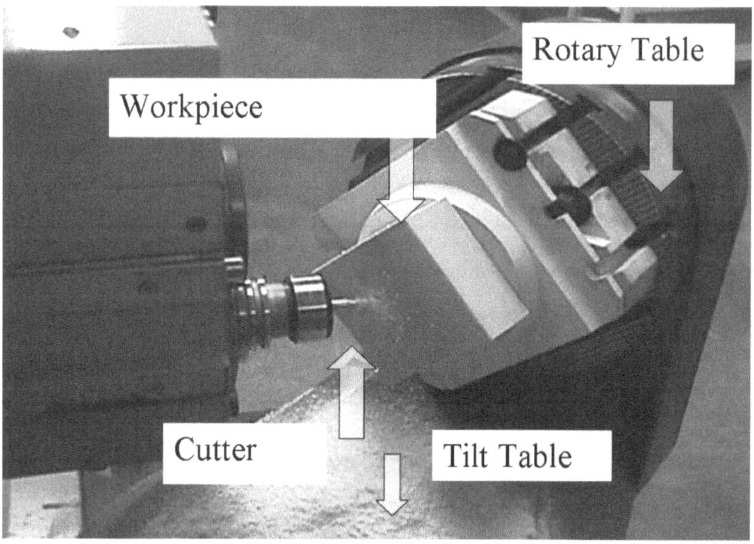

Fig. 2.1b. Workpiece, clumping device, and rotary tables of MAHO 600E

Fig. 2.1c. HERMLE UWF902H milling machine

The design of the part program takes into account the specific part geometry, the cutting process, the machine parameters and the cutting tools used. The program is fed to the machine controller that drives servomotors of the machine to move the table or the spindle by specified amounts in X, Y or Z and to tilt the cutter relative to the workpiece in one or two rotary axes. The part is machined by moving the cutter along a pre-computed path in space called the *tool path*. Generation of a part program to cut a specific part typically involves the following steps:

- Generation of the tool path in the workpiece coordinate system. Usually this step requires that a successive set of coordinates $W = (x_w, y_w, z_w)$ called cutter contact[2] (CC) points and the tool orientations $I = (I_x, I_y, I_z)$ in the workpiece coordinate system are distributed along a set of curves following a specified pattern such as the zigzag or the spiral curves (Fig. 2.2).
- The CL data are then computed from the set of CC points. The method for CC-CL data conversion must include the geometry of the cutting tool and the orientation of the tool relative to the workpiece. Various optimization techniques could be used at this step to generate optimal cutter location data (see, for instance, [2]).
- Each CL point is then transformed into the machine axial command $\Pi = (M, \mathfrak{R}) \in \mathbb{R}^5$, carrying three Cartesian coordinates $M = (x_m, y_m, z_m)$

[2] A cutter contact point refers to the coordinates where the tool cutting edge makes contact with the surface.

of the tool tip in the machine coordinate system and two rotation angles $\mathfrak{R} = (a, b)$. Converting coordinates of the CL points from the workpiece coordinate system to the machine coordinate system is called postprocessing. The postprocessor requires knowledge of the machine kinematics and the machine configuration.

- The axial commands are then converted into an NC part program following the format understood by the machine controller.

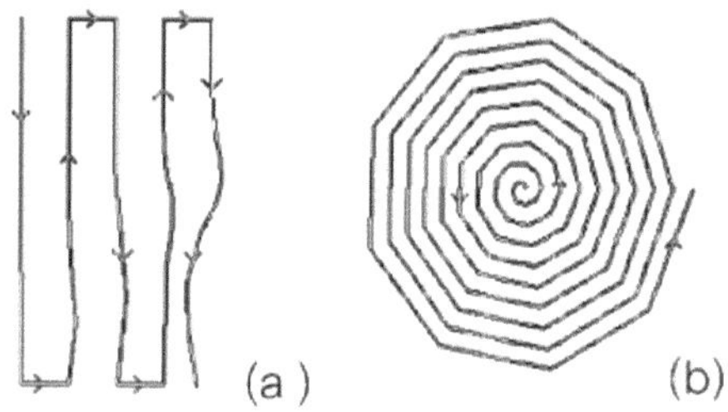

Fig. 2.2. Zigzag (**a**) and spiral (**b**) tool paths

The above sequence may or may not include an optimization and verification steps. The optimization step involves a cost function representing a certain type of error or a combination of several types of errors. The tool path is then modified or reconstructed entirely in such a way that the cost function is minimized or at least decreased.

The verification stage includes an actual machining or computer modeling of the material removal which produces the output as a solid model. The solid model can then be compared with the desired part and the efficiency of the optimization strategies can be evaluated.

2.2 NC Part Programming

The NC part program is a detailed plan (program) of manufacturing consisting of instructions called the NC blocks. The syntax of each instruction follows a specified format standardized by the ISO. This standard is maintained by many machine control units (MCU) manufacturers with some minor variations. When the machine is operated, the NC program is executed sequentially, one command or NC block at a time. An NC block is made up of words which consist of characters and digits. Each word begins with an address code which is a single letter character (A-Z) that defines what the computer should do with the numerical data that follows. This command structure is called the word address code. For example, the block

```
N20 G01 X100 Y-150 Z10.5 A30 B-10 F100 M03 S600
```

consists of 10 words: N20, G01, X100, Y-150, Z10.5, A30, B-10, F100, M03, and S600. The block N20 instructs the machine to move the tool from its current position to position $(x_m, y_m, z_m) = (100, -150, 10.5)$ linearly (G01) and to rotate the A-axis by 30 degrees and the B-axis by -10 degrees with a feedrate of 100 mm/min (F100) and with the spindle rotating clockwise (M03) at 600 rpm (S600).

The NC part program is usually written in a high level language such as APT (Automatic Programmed Tools) [7], UNIAPT[3], etc. It is then converted into the word address codes. Some commercial CAD/CAM systems such as Unigraphics, Pro Engineer, etc. include functions that can generate the word address codes directly from the tool paths.

A complete NC part program starts with a percent sign (%) followed by NC blocks represented in the ISO word address format. Each NC block starts with the block number indicated by an N word address. The block can contain one word address or a sequence of word addresses. The standardization of the English alphabet used to code the word addresses is given in Table 2.1.

Each letter other than G and M has a unique function and is followed by a parameter which is either an integer or a floating point number. The format of the ISO word address block is as follows:

```
N5 G2 X±53 Y±53 Z±53 U..V..W..I..J..K.. F5 S4 T4 M2 *
```

All word addresses except N are optional and the block format is order-sensitive. The numerical value after each word address indicates the maximum number of digits allowed for that particular word address. A single digit indicates an integer value and two digits indicate a real value. The first and the second digit indicate the maximum number of digits before and after the decimal point, respectively. For example, N5 means the block sequence number starts with the letter N followed by at most 5 digits (N0-N99999). X±53 means that the primary X motion dimension starts with the letter X followed by plus or minus sign and a real value written with at most 5 digits before the decimal point and at most 3 digits after the decimal point. The block ends with the symbol "*" or ";" or "carriage return" or "line feed". The usage of the word addresses for performing machine operation is given below.

Preparatory Functions

A pre-set function which assigns a type of the tool movement is indicated by a G word address followed by a two digit integer ranging from 00 to 99. Examples of machine tool action instructed by the G-codes are:

- Rapid move.

[3] UNIAPT is a minicomputer-based version of an Automatic Programmed Tool (APT) developed by a software company called United Computing to compute tool paths.

Table 2.1. List of word address codes

Character	Address for
A	Angular dimension around X axis
B	Angular dimension around Y axis
C	Angular dimension around Z axis
D	Angular dimension around special axis or third feed function*
E	Angular dimension around special axis or second feed function*
F	Feed function
G	Preparatory function
H	Unassigned
I	Distance to arc center or thread lead parallel to X
J	Distance to arc center or thread lead parallel to Y
K	Distance to arc center or thread lead parallel to Z
L	Unassigned
M	Miscellaneous function
N	Sequence number
O	Reference rewind stop
P	Third rapid traverse or tertiary motion dimension parallel to X*
Q	Third rapid traverse or tertiary motion dimension parallel to Y*
R	Third rapid traverse or tertiary motion dimension parallel to Z*
S	Spindle speed function
T	Tool function
U	Secondary motion dimension parallel to X*
V	Secondary motion dimension parallel to Y*
W	Secondary motion dimension parallel to Z*
X	Primary X motion dimension
Y	Primary Y motion dimension
Z	Primary Z motion dimension

* The letter may be used for other operations

- Controlled feed move in straight line or arc.
- Series of controlled feed moves that would result in a hole being bored, a workpiece cut (routed) to a specific dimension, or a decorative profile shape added to the edge of a workpiece.
- Set tool information such as offset.

The ISO standardized G-codes are given in Table 2.2.

Coordinate Functions

The coordinate of the tool tip are specified using a word address such as X, Y, Z, U, V, W, I, J, K, etc. These word addresses are normally signed real values.

Table 2.2. List of G-code functions

Code	Function
G00	Point-to-point positioning, rapid traverse
G01	Line interpolation
G02	Circular interpolation, clockwise (WC)
G03	Circular interpolation, anti-clockwise (CCW)
G04	Dwell
G05	Hold/Delay
G06	Parabolic interpolation
G07	Unassigned
G08	Acceleration of feedrate
G09	Deceleration of feedrate
G10	Linear interpolation for "long dimensions" (10 inches-100 inches)
G11	Linear interpolation for "short dimensions" (up to 10 inches)
G12	Unassigned
G13-G16	Axis designation
G17	XY plane designation
G18	ZX plane designation
G19	YZ plane designation
G20	Circular interpolation, CW for "long dimensions"
G21	Circular interpolation, CW for "short dimensions"
G22-G29	Unassigned
G30	Circular interpolation, CCW for "long dimensions"
G31	Circular interpolation, CCW for "short dimensions"
G32	Unassigned
G33	Thread cutting, constant lead
G34	Thread cutting, linearly increasing lead
G35	Thread cutting, linearly decreasing lead
G36-G39	Unassigned
G40	Cutter compensation-cancels to zero
G41	Cutter radius compensation-offset left
G42	Cutter radius compensation-offset right
G43	Cutter compensation-positive
G44	Cutter compensation-negative
G45-G52	Unassigned
G53	Deletion of zero offset
G54-G59	Datum point/zero shift
G60	Target value, positioning tolerance 1
G61	Target value, positioning tolerance 2 or loop cycle
G62	Rapid traverse positioning
G63	Tapping cycle
G64	Change in feedrate or speed
G65-G69	Unassigned
G70	Dimensioning in inch units
G71	Dimensioning in metric units
G72-G79	Unassigned
G80	Canned cycle cancelled
G81-G89	Canned drilling and boring cycles

Table 2.2. List of G-code functions (cont.)

Code	Function
G90	Specifies absolute input dimensions
G91	Specifies incremental input dimensions
G92	Programmed reference point shift
G93	Unassigned
G94	feedrate/min (inch units when combined with G70)
G95	feedrate/rev (metric units when combined with G71)
G96	Spindle feedrate for constant surface feed
G97	Spindle speed in revolution per minute
G98-G99	Unassigned

Feed Functions

The feedrate is specified by an F word address. The feedrate units are set by using an appropriate G-code (G94 or G95). The feed function is modal which means that once the feedrate is set it affects all the subsequent blocks until it is replaced by another F value.

Speed Function

The spindle speed is assigned by an S word address given in revolutions per minute (rpm) when used with the preparatory code G97. For a constant surface speed (G96) the spindle speed is given in meters per minute or feet per minute.

Tool Function

For the NC machines equipped with tool magazines and automatic tool changers, the tool selection is performed by a T word address followed by a number indicating the position of the tool in the magazine. For example T19 mean the tool stored in position 19 in the magazine is to be used. If the tool change is to be done manually, the T word address would signal the machine to stop.

Miscellaneous Function

Control of the clumping device, spindle, coolant and some other global operations is performed by an M word address. The standard M-codes are given in Table 2.3.

It must be noted that M80-M99 are unassigned which means that the MCU manufacturers can use those remaining for other purposes such as to accommodate new CNC programming capabilities, for example, real-time surface interpolation [6] or Pythagorean-hodograph curves [3–5].

Table 2.3. List of M-code functions

Code	Function
M00	Program stop, spindle and coolant off
M01	Optional programmable stop
M02	End of program-often interchangeable with M30
M03	Spindle on, CW
M04	Spindle on, CCW
M05	Spindle stop
M06	Tool change
M07	Coolant supply No. 1 on
M08	Coolant supply No. 2 on
M09	Coolant off
M10	Clamp
M11	Unclamp
M12	Unassigned
M13	Spindle on, CW + coolant on
M14	Spindle on, CCW + coolant on
M15	Rapid traverse in + direction
M16	Rapid traverse in - direction
M17-M18	Unassigned
M19	Spindle stop at specified angular position
M20-M29	Unassigned
M30	Program stop at end of tape + tape rewind
M31	Interlock by-pass
M32-M35	Constant cutting velocity
M36-M39	Unassigned
M40-M45	Gear changes; otherwise unassigned
M46-M49	Unassigned
M50	Coolant supply No. 3 on
M51	Coolant supply No. 4 on
M52-M54	Unassigned
M55	Linear cutter offset No. 1 shift
M56	Linear cutter offset No. 2 shift
M57-M59	Unassigned
M60	Piece part change
M61	Linear piece part shift, location 1
M62	Linear piece part shift, location 2
M63-M67	Unassigned
M68	Clamp piece part
M69	Unclamp piece part
M70	Unassigned
M71	Angular piece part shift, location 1
M72	Angular piece part shift, location 2
M73-M77	Unassigned
M78	Clamp non-activated machine bed-ways
M79	Unclamp non-activated machine bed-ways
M80-M99	Unassigned

Example of a simple NC milling program

We conclude this section with a simple NC program to cut a small rectangle 2 cm × 4 cm.

```
N1 X0 Y0 T01
N2 X0 Y40
N3 X20 Y40
N4 X20 Y0
N5 X0 Y0
N6 M00
```

Line 1 (N1) tells the machine to traverse to grid point $(X, Y) = (0, 0)$ and to use tool #1.
Line 2 tells the machine to traverse to grid point $(X, Y) = (0, 40)$.
Line 3 tells the machine to travel to grid point $(X, Y) = (20, 40)$.
Line 4 tells the machine to travel to grid point $(X, Y) = (20, 0)$.
Line 5 returns the machine to origin $(X, Y) = (0, 0)$.
Line 6 stops the machine.

2.3 Classification of Five-Axis Machines

A typical five-axis machine has three translational axes and two rotary axes. The five degrees of freedom is the minimum needed for establishing an arbitrary position and orientation of the cutting tool relative to the workpiece [1].

Five-axis machines can be classified into four groups based on the number of translational and rotary axes: (i) three translational axes and two rotary axes; (ii) two translational axes and three rotary axes; (iii) one translational axes and four rotary axes and (iv) five rotary axes. However, the majority of existing five-axis machine tools falls in the first group and all methods presented in this book are developed for this type of five-axis machines.

The machine axes can be assigned to the tool and to the machine tables using a variety of combinations. The kinematics chain diagram of the machine gives an idea of how this assignment has been done. For example, the kinematics chain diagram in Fig. 2.3 corresponds to the five-axis machine in Fig. 2.1a. The diagram tells us that the table (actually, the table is composed of two smaller connected tables) is carried by four axes including the both rotary axes and the tool is carried by only one axis.

Another useful classification of five-axis machines is based on the number of table and tool carrying axes. Finally, the machines can be also classified by the location of the rotary axes (on the tool or the table).

The above classifications of the five-axis machines are given below:

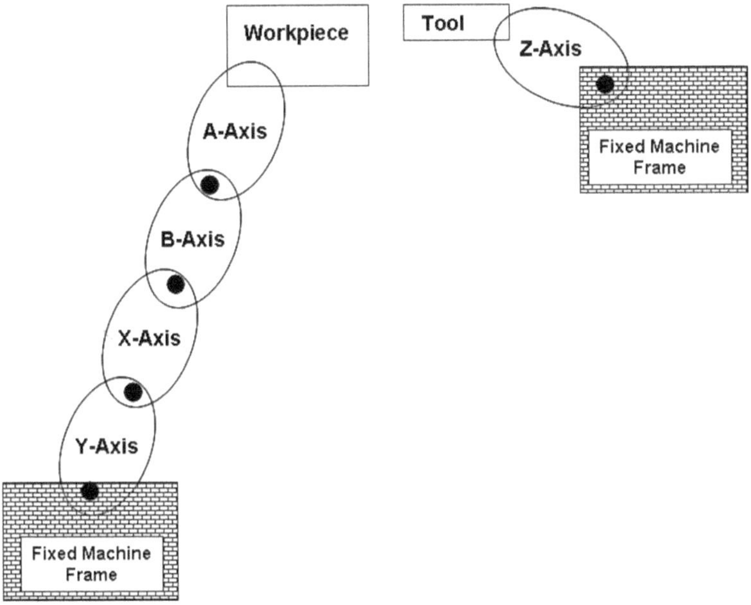

Fig. 2.3. Kinematic chain diagram of machine in Fig. 2.1a

Classification of five-axis machines based on the number of workpiece and tool carrying axes

1. *5/0 machine.* All axes carry the table and the tool is fixed in space. Since the tool is fixed, it is best used when the size of the workpiece is small.
2. *4/1 machine.* There are four table carrying axes and one tool carrying axis (Fig. 2.4).
3. *3/2 machine.* There are three table carrying axes and two tool carrying axes (Fig. 2.5).
4. *2/3 machine.* There are two table carrying axes and three tool carrying axes (Fig. 2.6). This configuration allows machining of large workpieces.
5. *1/4 machine.* There are one table carrying axis and four tool carrying axes.
6. *0/5 machine.* All axes carry the tool and the table is fixed in space. The configuration was designed to handle very heavy workpieces. However, a large number of links in the tool carrying kinematics chain could result in a considerable error due to elastic deformations in the axis slides.

Classification of five-axis machines based on the location of the two rotary axes

1. *2-0 machine.* The two rotary axes carry the table (Fig. 2.4).

2. *1-1 machine.* One rotary axis carries the table and one carries the tool (Fig. 2.5).
3. *0-2 machine.* The two rotary axes carry the tool (Fig. 2.6).

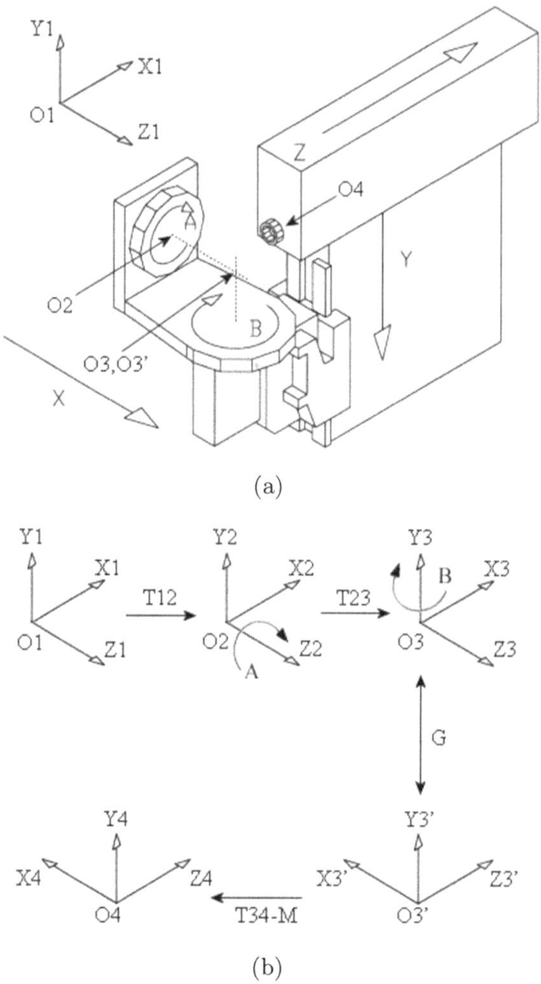

Fig. 2.4. (**a**) example of 2-0 machine and (**b**) the reference coordinate systems

Selection of a five-axis machine for a certain application is not an easy task. It can be done by analyzing characteristics of the workpiece such as the weight and the size. For example, a very heavy workpiece requires a machine with a horizontal table which makes it more convenient to fix and to handle the workpiece. Also, the machine should have a short workpiece kinematics chain since a heavy workpiece could result in an elastic deformations in the

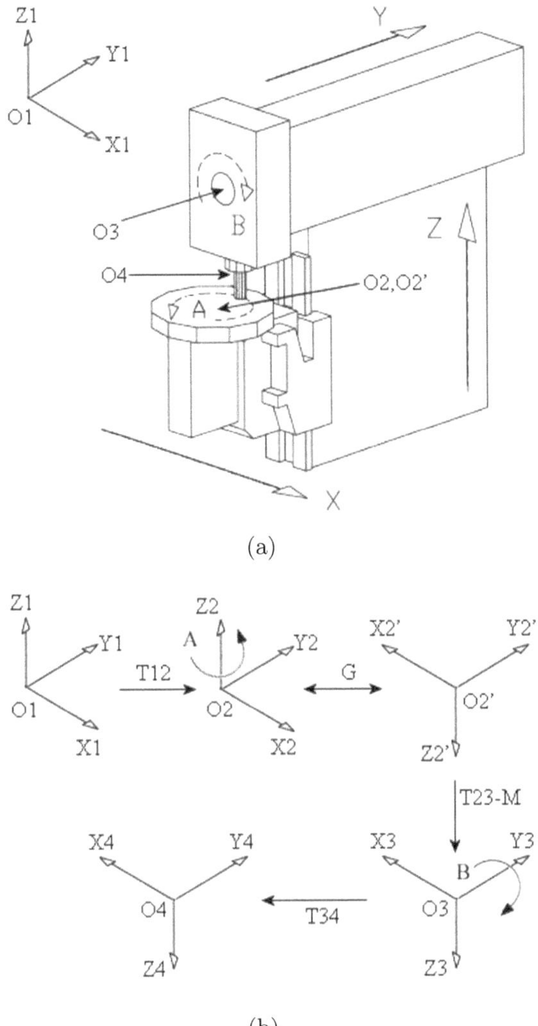

Fig. 2.5. (**a**) example of 1-1 machine and (**b**) the reference coordinate systems

axis slides that carry the table [1]. The suitability of the machine can be also analyzed in the framework of the kinematics error theory. The theoretical background for such analysis is given in Chap. 7 of this book.

2.4 Five-Axis Machine Kinematics

Five-axis machines offer five degrees of freedom which allow the cutting tool to incline or to tilt with regard to the workpiece while traveling from one point to

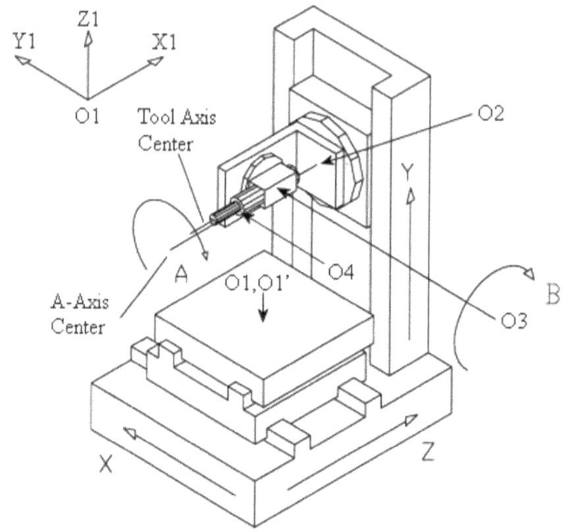

(a)

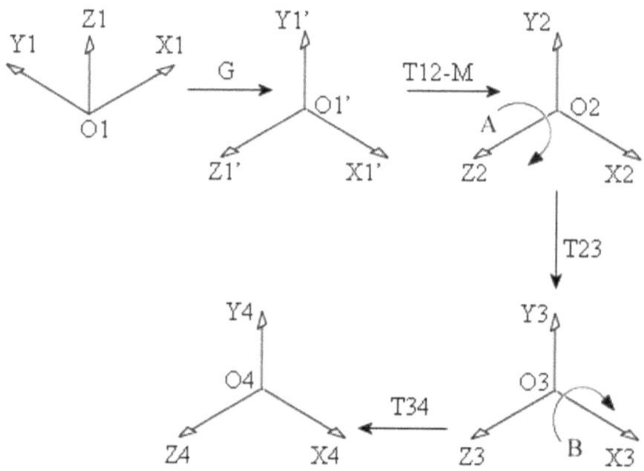

(b)

Fig. 2.6. (**a**) example of 0-2 machine and (**b**) the reference coordinate systems

another. Consequently, the tool tip affected by the machine kinematics travels in the workpiece coordinate system along a nonlinear trajectory. While this nonlinearity of the machine tools allows for complex surface parts to be milled it can also result in undercuts or overcuts or even in collisions between the cutting device and other parts of the machine.

The generation of NC part program that specifies the traveling path of the cutting device usually starts with the distribution of CC points on the surface. In five-axis machining, each CC point is also associated with a tool vector positioned in a right-handed, rectangular coordinate system. CC points are then converted into CL points which specify the coordinates of the tool tip in the workpiece coordinate systems. Finally, the CL points are transformed into the machine coordinates (or G-Codes) to control the motions of the machine and finally into the NC program.

The conversion from a CC point to a CL point depends on the tool geometry and inclination (see an example given in Chap. 3). The transformation of the CL points from the workpiece coordinate system to the machine coordinate system for a particular five-axis machine requires information regarding the kinematics of that machine.

The kinematics $\mathfrak{K} \equiv \mathfrak{K}(\mathfrak{R}, M)$ of a machine is a transformation from a point $M = (x_m, y_m, z_m)$ in the machine coordinates to a point $W = (x_w, y_w, x_w)$ in the workpiece coordinates where $\mathfrak{R} = (a, b)$ is a pair of rotation angles. For simplicity, the transformation will be denoted $\mathfrak{K}(M)$ when possible. Also, let $I = (I_x, I_y, I_z)$ denote the tool vector orientation specified in the workpiece coordinate system.

To derive an equation of the machine kinematics, the following coordinate systems are introduced: the workpiece coordinate system O_1, the coordinate system of the first rotary part O_2, the coordinate system of the second rotary part O_3, and the coordinate system of the spindle O_4 (see Figs. 2.4-2.6). The first rotary axis is called the A-axis, whereas the second rotary axis is called the B-axis. When an NC machine is powered up, the machine's measuring system is initialized by traveling the tool of the machine to the reference point in the machine coordinate system so that the machine control knows the absolute position of each slide.

Consider a five-axis machine displayed in Fig. 2.1a. The machine is of the 2-0 type. The two rotary axes A and B carry the table. The motions of the machine slides from reference position in machine coordinate system are as follows (the onlooker standing in front of the machine looking in the positive Z machine direction):

- x_m plus: the table moves to the right
- y_m plus: the table moves downwards
- z_m plus: the tool moves away from the viewer
- a plus: the table carrying workpiece rotates clockwise around the A-axis
- b plus: the table swings from right to left around the B-axis

The positive direction of each Cartesian axis slide is selected such that the cutting device appears to move away from the workpiece in the positive direction of the corresponding axis in the spindle coordinate system O_4 (see Fig. 2.4). In other words, the positive direction of an axis slide that carries the table is in the opposite direction of the positive direction of the corresponding axis in O_4 and the positive direction of an axis slide that carries the tool is in the same direction of the positive direction of the corresponding axis in O_4. For example, the slides X and Y of the 2-0 machine in Fig. 2.4 which carry the table have positive directions that are in opposite direction of positive X_4

and Y_4, respectively, while slide Z which carries the tool has positive direction that is in the same direction of positive Z_4.

The transformation from the workpiece to the machine coordinates consists of three types of transformations: coordinate transformation to another reference system without motions and transformation corresponding to actual machine rotations, and actual translations. The 2-0 machine (Fig. 2.4) kinematics equations are derived as follows:

Step 1: Coordinate transformation $O_1 \to O_2$

Point W in O_1 is transformed into point P_2 in O_2.

$$\boxed{P_2 = W + T_{12},}$$

where T_{12} is the coordinate of the center of O_1 in O_2.

Step 2: Rotation around A-axis in O_2 by a

Point P_2 is rotated around A-axis in O_2 by a to a new location P_{2A}.

$$\boxed{P_{2A} = A[a]P_2 = A[a](W + T_{12}),}$$

where $A[a] = \begin{bmatrix} \cos a & \sin a & 0 \\ -\sin a & \cos a & 0 \\ 0 & 0 & 1 \end{bmatrix}$ is the rotation matrix around A-axis.

Step 3: Coordinate transformation $O_2 \to O_3$

Point P_{2A} in O_2 is transformed into point P_3 in O_3.

$$\boxed{P_3 = P_{2A} + T_{23} = A[a](W + T_{12}) + T_{23},}$$

where T_{23} is the coordinate of the center of O_2 in O_3.

Step 4: Rotation around B-axis in O_3 by b

Point P_3 is rotated around B-axis in O_3 by b to a new location P_{3B}.

$$\boxed{P_{3B} = B[b]P_3 = B[b](A[a](W + T_{12}) + T_{23}),}$$

where $B[b] = \begin{bmatrix} \cos b & 0 & -\sin b \\ 0 & 1 & 0 \\ \sin b & 0 & \cos b \end{bmatrix}$ is the rotation matrix around B-axis.

Step 5: Coordinate rotation (axis alignment) $O_3 \to O_3'$

We now perform axis alignment for O_3 (the last reference coordinate system on the table) to O_4 (the first reference coordinate system on the tool).

$$\boxed{P_3' = GP_{3B} = GB[b](A[a](W + T_{12}) + T_{23}),}$$

where $G = \begin{bmatrix} 0 & 0 & -1 \\ 0 & 1 & 0 \\ 1 & 0 & 0 \end{bmatrix}$ is the axis alignment matrix.

Point P_3' is now given in the coordinate system O_3' where each direction of the three Cartesian coodinate axes lies in the same direction of the corresponding Cartesian coordinate axis of O_4

Step 6: Coordinate transformation $O_3' \to O_4$ with machine slide translation M

Point P_3' in O_3' is transformed to point P_4 in O_4. Recall that the workpiece attached to the table will appear to move away from the spindle in the negative directions in O_4 for positive value of machine slide translation M.

Thus for machines slide translation M, the center of O'_3 will be at location $T_{34} - M$ in O_4 where T_{34} is the coordinate of O'_3 in O_4 with respect to the machine zero point $M = (0,0,0)$. Finally, the transformation of the point P'_3 in O'_3 to point P_4 in O_4 is given as follows:

$$\boxed{P_4 = P'_3 + T_{34} - M = GB[b](A[a](W + T_{12}) + T_{23}) + T_{34} - M.}$$

Let $T_4 = (0,0,-L)$ denote the coordinate of the tool tip in coordinate system O_4 where L is the tool lenght. Equating P_4 and T_4 yields

$$T_4 = P_4 = GB[b]\left(A[a]\left(W + T_{12}\right) + T_{23}\right) + T_{34} - M. \tag{2.1}$$

After rearrangement,

$$\begin{aligned} M &= GB[b]\left(A[a]\left(W + T_{12}\right) + T_{23}\right) + T_{34} - T_4, \\ W &= A^{-1}[a]\left(B^{-1}[b]G^{-1}\left(M - T_{34} + T_4\right) - T_{23}\right) - T_{12}, \end{aligned} \tag{2.2}$$

The equation relating the two rotation angles a and b and the tool vector orientation $I = (I_x, I_y, I_z)$ can be derived as follows. Let T'_1 be the coordinate of $T'_4 = (0,0,-L+1)$ in O_1 and is given by

$$T'_1 = A^{-1}[a]\left(B^{-1}[b]G^{-1}\left(M - T_{34} + T'_4\right) - T_{23}\right) - T_{12}. \tag{2.3}$$

The unit vector $T'_4 - T_4 = (0,0,1)$ is equivalent to the tool vector orientation and is expressed in O_1 coordinate system as follows:

$$I = T'_1 - W = A^{-1}[a]B^{-1}[b]G^{-1}\left(T'_4 - T_4\right) = \begin{bmatrix} \cos(a)\cos(b) \\ \sin(a)\cos(b) \\ -\sin(b) \end{bmatrix}. \tag{2.4}$$

Inverting with regard to a and b yields

$$\begin{aligned} a &= \begin{cases} \tan^{-1}\left(\dfrac{I_y}{I_x}\right) & \text{if } I_x > 0 \text{ and } I_y \geq 0, \\ \tan^{-1}\left(\dfrac{I_y}{I_x}\right) + \pi & \text{if } I_x < 0, \\ \tan^{-1}\left(\dfrac{I_y}{I_x}\right) + 2\pi & \text{otherwise,} \end{cases} \\ b &= -\sin I_z, \end{aligned} \tag{2.5}$$

Derivations of machine kinematics equations for arbitrary configurations of five-axis machines can be done in a similar fashion. For the 1-1 machine shown in Fig. 2.5, the coordinate of the tool tip in O_4 is given by

$$T_4 = B[b]\left(GA[a]\left(W + T_{12}\right) + T_{23} - M\right) + T_{34}. \tag{2.6}$$

Clearly,

$$
\begin{aligned}
&M = GA[a]\left(W + T_{12}\right) + T_{23} + B^{-1}[b]\left(T_{34} - T_4\right),\\
&W = A^{-1}[a]G^{-1}\left(M - T_{23} - B^{-1}[b]\left(T_{34} - T_4\right)\right) - T_{12},\\
&I = T_1' - W = A^{-1}[a]G^{-1}B^{-1}[b]\left(T_4' - T_4\right) = \begin{bmatrix} \cos(a)\sin(b) \\ -\sin(a)\sin(b) \\ \cos(b) \end{bmatrix},\\
&a = \begin{cases} -\tan^{-1}\left(\dfrac{I_y}{I_x}\right) & \text{if } I_x > 0 \text{ and } I_y \le 0,\\ -\tan^{-1}\left(\dfrac{I_y}{I_x}\right) + \pi & \text{if } I_x < 0,\\ -\tan^{-1}\left(\dfrac{I_y}{I_x}\right) + 2\pi & \text{otherwise,} \end{cases}\\
&b = \cos^{-1} I_z,
\end{aligned}
\tag{2.7}
$$

where

$$
\begin{aligned}
A[a] &= \begin{bmatrix} \cos a & -\sin a & 0 \\ \sin a & \cos a & 0 \\ 0 & 0 & 1 \end{bmatrix},\\
B[b] &= \begin{bmatrix} \cos b & 0 & -\sin b \\ 0 & 1 & 0 \\ \sin b & 0 & \cos b \end{bmatrix},\\
G &= \begin{bmatrix} -1 & 0 & 0 \\ 0 & 1 & 0 \\ 0 & 0 & -1 \end{bmatrix},\\
T_4 &= (0, 0, L).
\end{aligned}
\tag{2.8}
$$

For the 0-2 machine shown in Fig. 2.6, the coordinate of the tool tip T_4 in O_4 is given by

$$
T_4 = B[b]\left(A[a]\left(GW + T_{12} - M\right) + T_{23}\right) + T_{34}. \tag{2.9}
$$

Clearly,

$$
\begin{aligned}
&M = GW + T_{12} + A^{-1}[a]\left(T_{23} + B^{-1}[b]\left(T_{34} - T_4\right)\right),\\
&W = G^{-1}\left(M - T_{12} - A^{-1}[a]\left(T_{23} + B^{-1}[b]\left(T_{34} - T_4\right)\right)\right),\\
&I = T_1' - W = G^{-1}A^{-1}[a]B^{-1}[b]\left(T_4' - T_4\right) = \begin{bmatrix} \cos(b) \\ \sin(a)\sin(b) \\ -\cos(a)\sin(b) \end{bmatrix},\\
&a = \begin{cases} -\tan^{-1}\left(\dfrac{I_y}{I_z}\right) & \text{if } I_y \le 0 \text{ and } I_z > 0,\\ -\tan^{-1}\left(\dfrac{I_y}{I_z}\right) + \pi & \text{if } I_z < 0,\\ -\tan^{-1}\left(\dfrac{I_y}{I_z}\right) + 2\pi & \text{otherwise,} \end{cases}\\
&b = -\cos^{-1} I_x,
\end{aligned}
\tag{2.10}
$$

where

$$
\begin{aligned}
A[a] &= \begin{bmatrix} \cos a & -\sin a & 0 \\ \sin a & \cos a & 0 \\ 0 & 0 & 1 \end{bmatrix}, \\
B[b] &= \begin{bmatrix} 1 & 0 & 0 \\ 0 & \cos b & -\sin b \\ 0 & \sin b & \cos b \end{bmatrix}, \\
G &= \begin{bmatrix} 0 & -1 & 0 \\ 0 & 0 & 1 \\ -1 & 0 & 0 \end{bmatrix}, \\
T_4 &= (0, 0, L).
\end{aligned}
\tag{2.11}
$$

2.5 Five-Axis Machining Example

Creating objects such as industrial molds and dies by five-axis machines requires a computer program that instructs the machine to move the cutting device through a block of a raw material in such a way that it removes all the excess material, leaving behind the desired shape. The path of the tool tip is called the tool path and the process of creating the tool path is called the tool path generation.

Consider a surface $S(u, v) = (x(u, v), y(u, v), z(u, v))$ described by a set of parametric equations given by

$$
\begin{aligned}
x &= 100u - 50, \\
y &= 100v - 50, \\
z &= 10((u + 0.5)^2 + v^2) - 50,
\end{aligned}
\tag{2.12}
$$

where $0 \le u \le 1$ and $0 \le v \le 1$. The surface plot is shown in Fig. 2.7.

A simple way to generate the tool path for five-axis machining of this surface is the zigzag pattern shown in Fig. 2.2a. There are 10 tool tracks equally spaced in the v-direction. In turn, each track contains 10 CC points equally spaced in the u-direction Each CC point is associated with a tool vector aligned with the surface normal vector. The method of calculating the surface normal vector can be found in Sect. 3.1. The plot of the CC points along with the tool vector orientations is shown in Fig. 2.8. Since the tool vector orientation is aligned with the surface normal vector, the cutter contact point is assumed to coincide with the tool tip. Thus, the CL points are set to have the same location as the CC points.

Next, CL points are converted into the machine commands or the G-codes. For MAHO machine shown in Fig. 2.1a, (2.2) and (2.5) are used to compute the machine coordinates M and the two rotation angles, a and b. The machine parameters are given below:

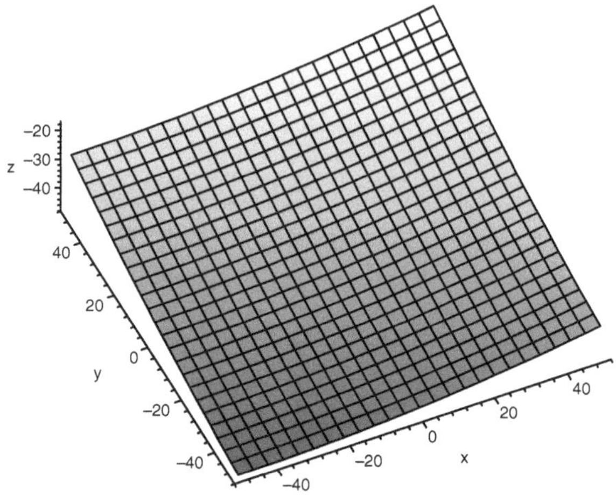

Fig. 2.7. Simple shape parametric surface

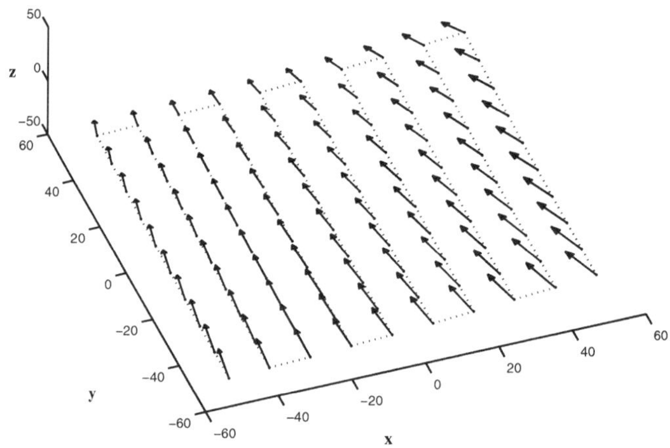

Fig. 2.8. Zigzag tool path for surface (2.12) (Fig. 2.7)

$$
\begin{aligned}
T_{12} &= (0.0, 0.0, 213.58),\\
T_{23} &= (0.08, 0.0, -250.314),\\
T_{34} &= (0.0, 0.0, -350.0),\\
L &= 177.651.
\end{aligned}
$$

The coordinates of the CL points in the workpiece and the machine coordinate systems are partially listed in Table 2.4. The G-codes are straightforwardly derived from the machine coordinates. x_m, y_m and z_m are prefixed with the address words X, Y and Z, respectively. Likewise, the two rotation angles are prefixed with the address words A and B. Each CL point constitutes one block of the G-code. The G-code begins with the word address N followed by the block number. The G-codes are given in Table 2.5.

Table 2.4. List of CL points for tool path in Fig. 2.8

Point Number	x_w	y_w	z_w	I_x	I_y	I_z
1	-50.000	-50.000	-47.500	-0.100	0.000	0.995
2	-50.000	-38.889	-47.377	-0.099	-0.022	0.995
3	-50.000	-27.778	-47.006	-0.099	-0.044	0.994
4	-50.000	-16.667	-46.389	-0.099	-0.066	0.993
5	-50.000	-5.556	-45.525	-0.099	-0.088	0.991
6	-50.000	5.556	-44.414	-0.099	-0.110	0.989
7	-50.000	16.667	-43.056	-0.099	-0.132	0.986
8	-50.000	27.778	-41.451	-0.098	-0.153	0.983
9	-50.000	38.889	-39.599	-0.098	-0.174	0.980
10	-50.000	50.000	-37.500	-0.098	-0.195	0.976
11	-38.889	50.000	-36.265	-0.119	-0.195	0.974
12	-38.889	38.889	-38.364	-0.119	-0.174	0.978
13	-38.889	27.778	-40.216	-0.120	-0.153	0.981
14	-38.889	16.667	-41.821	-0.120	-0.131	0.984
15	-38.889	5.556	-43.179	-0.121	-0.110	0.987
16	-38.889	-5.556	-44.290	-0.121	-0.088	0.989
17	-38.889	-16.667	-45.154	-0.121	-0.066	0.990
18	-38.889	-27.778	-45.772	-0.121	-0.044	0.992
19	-38.889	-38.889	-46.142	-0.121	-0.022	0.992
20	-38.889	-50.000	-46.265	-0.121	0.000	0.993
			⋮			
91	50.000	50.000	-17.500	-0.282	-0.188	0.941
92	50.000	38.889	-19.599	-0.283	-0.168	0.944
93	50.000	27.778	-21.451	-0.284	-0.147	0.947
94	50.000	16.667	-23.056	-0.285	-0.127	0.950
95	50.000	5.556	-24.414	-0.286	-0.106	0.952
96	50.000	-5.556	-25.525	-0.286	-0.085	0.954
97	50.000	-16.667	-26.389	-0.287	-0.064	0.956
98	50.000	-27.778	-27.006	-0.287	-0.043	0.957
99	50.000	-38.889	-27.377	-0.287	-0.021	0.958
100	50.000	-50.000	-27.500	-0.287	0.000	0.958

Figure 2.9 shows the result of cutting surface (2.12) (Fig. 2.7) using the 10×10 tool path. The simulation is done with Unigraphics 18. Clearly, the tool path contains insufficient number of CL points and left a great amount of an excess material.

Figure 2.10 displays the result of cutting using a 20×20 zigzag tool path. The tool now removes most of the excess material and produces a reasonable model of the desired surface.

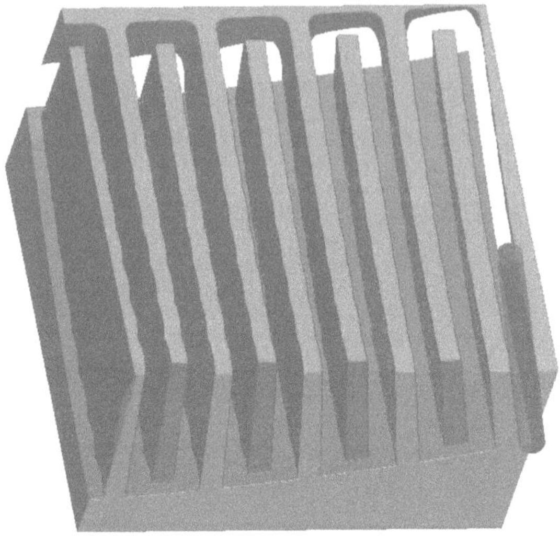

Fig. 2.9. Cutting simulation of surface (2.12) (Fig. 2.7) with 10×10 tool path

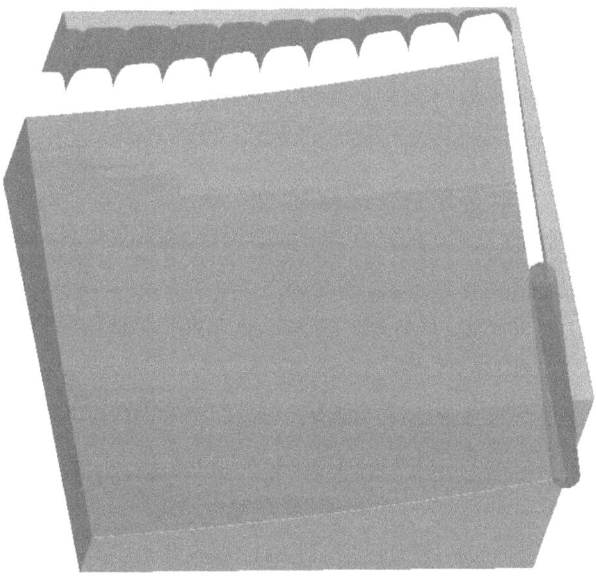

Fig. 2.10. Cutting simulation of surface (2.12) (Fig. 2.7) with 20×20 tool path

Table 2.5. G-codes for tool path in Fig. 2.8

Point Number	G-Code						
1	N10	G01	X58.213	Y50.000	Z-251.183	A180.000	B-84.289 F100
2	N11	G01	X65.599	Y27.116	Z-250.181	A192.529	B-84.151 F100
3	N12	G01	X65.823	Y5.077	Z-249.387	A203.962	B-83.755 F100
4	N13	G01	X60.483	Y-13.868	Z-248.802	A213.690	B-83.147 F100
5	N14	G01	X51.687	Y-29.066	Z-248.426	A221.634	B-82.379 F100
6	N15	G01	X41.073	Y-40.881	Z-248.259	A228.013	B-81.498 F100
7	N16	G01	X29.636	Y-50.000	Z-248.301	A233.130	B-80.538 F100
8	N17	G01	X17.907	Y-57.080	Z-248.549	A237.265	B-79.523 F100
9	N18	G01	X6.142	Y-62.644	Z-249.002	A240.642	B-78.471 F100
10	N19	G01	X-5.544	Y-67.082	Z-249.657	A243.435	B-77.396 F100
11	N20	G01	X-5.058	Y-59.256	Z-248.513	A238.570	B-76.809 F100
12	N21	G01	X6.126	Y-54.078	Z-247.854	A235.491	B-77.826 F100
13	N22	G01	X17.155	Y-47.741	Z-247.398	A231.843	B-78.810 F100
14	N23	G01	X27.830	Y-39.929	Z-247.146	A227.490	B-79.747 F100
15	N24	G01	X37.806	Y-30.270	Z-247.101	A222.274	B-80.621 F100
16	N25	G01	X46.515	Y-18.380	Z-247.264	A216.027	B-81.406 F100
17	N26	G01	X53.090	Y-3.990	Z-247.637	A208.610	B-82.074 F100
18	N27	G01	X56.376	Y12.815	Z-248.219	A199.983	B-82.590 F100
19	N28	G01	X55.170	Y31.305	Z-249.010	A190.305	B-82.919 F100
20	N29	G01	X48.751	Y50.000	Z-250.009	A180.000	B-83.032 F100
				⋮			
91	N100	G01	X-46.756	Y-13.868	Z-246.860	A213.690	B-70.173 F100
92	N101	G01	X-40.711	Y-7.966	Z-246.207	A210.651	B-70.775 F100
93	N102	G01	X-35.461	Y-1.644	Z-245.751	A207.408	B-71.328 F100
94	N103	G01	X-31.117	Y5.077	Z-245.495	A203.962	B-71.825 F100
95	N104	G01	X-27.787	Y12.156	Z-245.440	A200.323	B-72.260 F100
96	N105	G01	X-25.578	Y19.531	Z-245.588	A196.504	B-72.626 F100
97	N106	G01	X-24.580	Y27.116	Z-245.940	A192.529	B-72.917 F100
98	N107	G01	X-24.860	Y34.805	Z-246.496	A188.427	B-73.129 F100
99	N108	G01	X-26.453	Y42.476	Z-247.256	A184.236	B-73.258 F100
100	N109	G01	X-29.357	Y50.000	Z-248.219	A180.000	B-73.301 F100

The above method for tool path generation works fine for simple surfaces. However, for complex surfaces such as the one shown in Fig. 2.11, the method could produce, overcuts, undercuts at the CC points as well as nonlinear trajectory loops between the points which could destroy the workpiece or even damage the machine itself (see Figs. 2.12 and 2.13).

Furthermore, for the concave surfaces, the cutting tool needs to be inclined to avoid curvature interference. Besides, the distance between the tool tracks is fixed. Therefore, if the distance is too large, the tool leaves behind ridges of unremoved material which called cusps or scallops. On the other hand, a too small distance results in a redundant path when the tool makes unnecessary cuts in the same areas between the tracks.

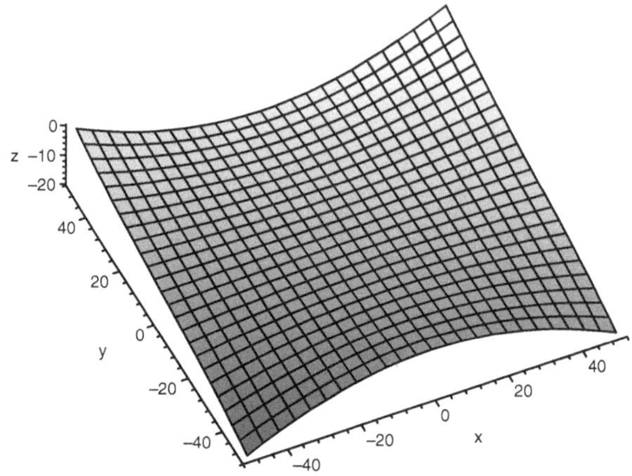

Fig. 2.11. Bezier surface

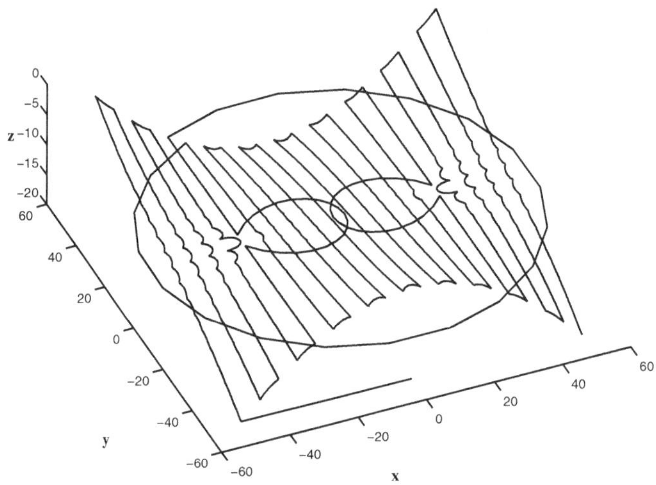

Fig. 2.12. Tool trajectories with loops of 20×20 tool path for surface in Fig. 2.11

References

[1] Bohez, E. L. J. 2002. Five-axis milling machine tool kinematic chain design and analysis. *International Journal of Machine Tools and Manufacture*, 42(4):505–520.

[2] Choi, B. K., Park, J. W., and Jun, C. S. 1993. Cutter-location data optimization in 5-axis surface machining. *Computer-Aided Design*, 25(6):377–386.

[3] Farouki, R. T., Manjunathaiah, J., Nicholas, D., Yuan, G.-F., and Jee, S. 1998. Variable-feedrate CNC interpolators for constant material re-

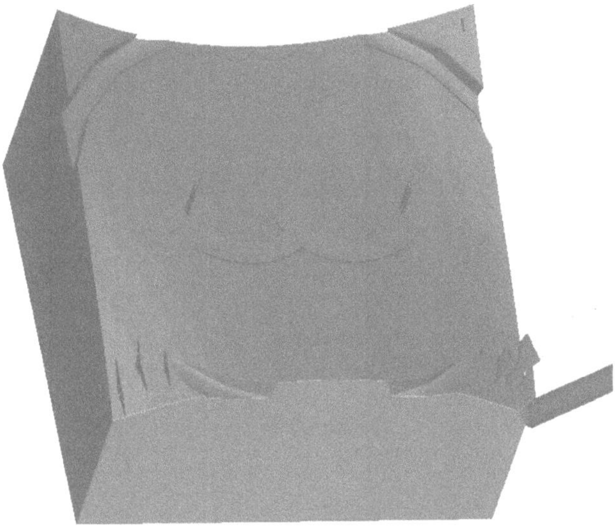

Fig. 2.13. Cutting simulation of surface in Fig. 2.11 with 20×20 tool path

moval rates along Pythagorean-hodograph curves. *Computer-Aided Design*, 30(8):631–640.

[4] Farouki, R. T., Manjunathaiah, J., and Yuan, G.-F. 1999. G codes for the specification of Pythagorean-hodograph tool paths and associated feedrate functions on open-architecture CNC machines. *International Journal of Machine Tools and Manufacture*, 39(1):123–142.

[5] Farouki, R. T. and Shah, S. 1996. Real-time CNC interpolators for Pythagorean-hodograph curves. *Computer Aided Geometric Design*, 13(7):583–600.

[6] Lin, R.-S. 2000. Real-time surface interpolator for 3-D parametric surface machining on 3-axis machine tools. *International Journal of Machine Tools and Manufacture*, 40(10):1513–1526.

[7] Ross, D. T. 1978. Origins of the APT language for automatically programmed tools. In *HOPL-1: The first ACM SIGPLAN conference on History of programming languages*, pages 61–99.

3

Fundamental Issues in Tool Path Planning

There are a number of issues involved in tool path planning for five-axis NC machining. Four fundamental issues are discussed in this chapter, namely, surface representation, machining strip width estimation, optimal tool orientation and forward step (kinematics) error.

3.1 Surface Representation

Five-axis NC machines are widely used to machine dies, molds, turbine blades, aerospace and automotive parts, etc. These parts usually have complex geometry and are represented by parametric surfaces. Representation of surface by parametric equations allows the simple evaluation of differential properties of the surface. Many tool path generation techniques require calculation of certain surface properties such as the surface normal vector and the normal curvature. The close-form formula for commonly used surface properties are given below.

Let $S \equiv S(u, v)$ be the required parametric surface, the *unit normal vector*, $\mathbf{n}$, is computed from the relation

$$\mathbf{n} = \frac{S_u \times S_v}{|S_u \times S_v|}. \tag{3.1}$$

The *first fundamental form* (or line element), $\mathbf{I}$, is defined by

$$\begin{aligned} \mathbf{I} &= dS \cdot dS, \\ &= (S_u du + S_v dv) \cdot (S_u du + S_v dv), \\ &= S_u \cdot S_u du^2 + 2S_u \cdot S_v dudv + S_v \cdot S_v dv^2, \\ &= Edu^2 + 2Fdudv + Gdv^2, \end{aligned} \tag{3.2}$$

where

$$E = S_u \cdot S_u, \tag{3.3}$$
$$F = S_u \cdot S_v, \tag{3.4}$$
$$G = S_v \cdot S_v. \tag{3.5}$$

The *second fundamental form*, **II**, is defined by

$$\begin{aligned}
\mathbf{II} &= -d\mathbf{n} \cdot dS, \\
&= -(\mathbf{n}_u du + \mathbf{n}_v dv) \cdot (S_u du + S_v dv), \\
&= -\mathbf{n}_u \cdot S_u du^2 - (\mathbf{n}_u \cdot S_v + \mathbf{n}_v \cdot S_u) dudv - \mathbf{n}_v \cdot S_v dv^2, \\
&= \mathbf{n} \cdot S_{uu} du^2 + (\mathbf{n} \cdot S_{vu} + \mathbf{n} \cdot S_{uv}) dudv + \mathbf{n} \cdot S_{vv} dv^2, \\
&= \mathbf{n} \cdot S_{uu} du^2 + 2\mathbf{n} \cdot S_{uv} dudv + \mathbf{n} \cdot S_{vv} dv^2, \\
&= edu^2 + 2f dudv + gdv^2,
\end{aligned} \tag{3.6}$$

where

$$e = \mathbf{n} \cdot S_{uu}, \tag{3.7}$$
$$f = \mathbf{n} \cdot S_{uv}, \tag{3.8}$$
$$g = \mathbf{n} \cdot S_{vv}. \tag{3.9}$$

The *normal curvature* of S in the direction $\mathbf{v} = aS_u + bS_v$, is given by

$$k(\mathbf{v}) = \frac{ea^2 + 2fab + gb^2}{Ea^2 + 2Fab + Gb^2}. \tag{3.10}$$

The *principal curvatures*, which are the maximum and minimum of the normal curvature, are given by

$$k_{\max} = H + \sqrt{H^2 - K}, \tag{3.11}$$
$$k_{\min} = H - \sqrt{H^2 - K}, \tag{3.12}$$

where K and H are the *Gaussian curvature* and the *mean curvature*, respectively. They are given by

$$K = \frac{eg - f^2}{EG - F^2} = k_{\min} k_{\max}, \tag{3.13}$$
$$H = \frac{eG - 2fF + gE}{2(EG - F^2)} = \frac{1}{2}(k_{\min} + k_{\max}). \tag{3.14}$$

The direction $\mathbf{v} = aS_u + bS_v$ associated with a principal curvature, k, is computed using the relations

$$(e - kE)a + (f - kF)b = 0, \tag{3.15}$$
$$(f - kF)a + (g - kG)b = 0. \tag{3.16}$$

The surface points are classified into six different types, depending on the values of K and H as follows.

Concave elliptic point If $K > 0$ and $H > 0$, the surface lies entirely on the surface normal side ($\mathbf{n}$) of the tangent plane in its neighborhood. Both the principal curvatures are positive.

Convex elliptic point If $K > 0$ and $H < 0$, the surface lies entirely on the opposite side ($-\mathbf{n}$) of the tangent plane in its neighborhood. Both the principal curvatures are negative.

Hyperbolic point If $K < 0$, the surface lies entirely on both sides of the tangent plane in its neighborhood. The principal curvatures have different signs.

Concave parabolic point If $K = 0$ and $H > 0$, the surface lies entirely on the surface normal side ($\mathbf{n}$) of the tangent plane in its neighborhood. One of the principal curvatures is positive and one is zero.

Convex parabolic point If $K = 0$ and $H < 0$, the surface lies entirely on the opposite side ($-\mathbf{n}$) of the tangent plane in its neighborhood. One of the principal curvatures is negative and one is zero.

Planar umbilic point If $K = 0$ and $H = 0$, the surface lies entirely in the tangent plane in the neighborhood of the umbilic point.

3.2 Machining Strip Width Estimation

In five-axis NC machining, several types of cutters can be used to mill the surface. The bottom of the cutter may be flat or rounded, as shown in Fig. 3.1. A ball-end cutter has rounded bottom where the filleted portion is the same size as the cutter radius. A rounded tool with a slight corner radius is called a toroidal cutter. A flat-end cutter has flat bottom. It can be used to remove the material much more effectively than the ball-end cutter but is more difficult to position relative to a sculptured surface. No matter which type of the milling tools is used, only a few shapes of surface can be cut exactly. This is because the shape of the tool almost never matches the shape of the surface. Irregular scallops between finishing tool passes are inevitably generated on the machined surface. The term scallop refers to ridges, cusps and other surface protrusions left between adjacent overlapping tool passes that extend above the design surface profile. In five-axis NC machining, controlling scallop height is a significant factor since a small scallop height significantly reduces the manual surface grinding and smoothing. Given the maximum scallop height h, the next adjacent tool path offset is found by computing the machining strip width at the current cutter contact point. Calculation of the machining strip width for various type of machining tools can be found in the literature (see, for instance, [2, 11, 12]). A simple but efficient algorithm for estimating the machining strip width of a flat-end cutter is presented here.

Machining strip width estimation for flat-end cutter

Let us first introduce a local coordinate system (O_l, $\mathbf{x}_l$, $\mathbf{y}_l$, $\mathbf{z}_l$) at the cutter contact point O_l as shown in Fig. 3.2. $\mathbf{x}_l$ denotes the normalized projection of

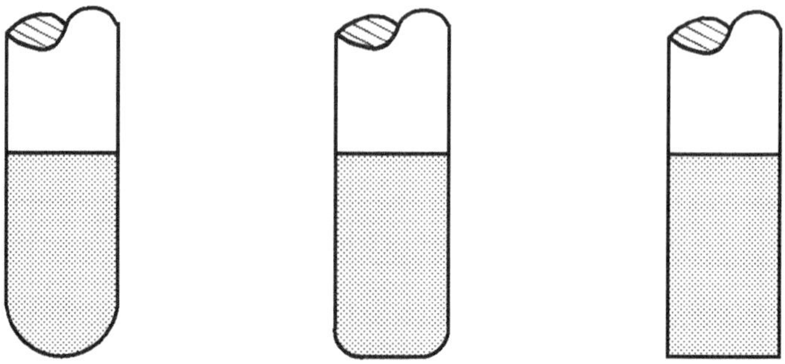

Fig. 3.1. Examples of cutting tools: (a) a ball-end cutter, (b) a toroidal cutter, (c) a flat-end cutter

the tool cutting direction[1] onto the tangent plane, $\mathbf{z}_l$ denotes the unit normal vector of the surface, and $\mathbf{y}_l = \mathbf{z}_l \times \mathbf{x}_l$. In order to avoid gouging, the tool is first rotated by an inclination angle λ about the $\mathbf{y}_l$ axis vector, then by a tilt angle ω about the $\mathbf{z}_l$ axis vector. The projected bottom edge of a flat-end cutter with radius r onto the $\mathbf{y}_l$-$\mathbf{z}_l$ plane becomes an ellipse, which is called the *effective cutting shape*.

A point P on the bottom edge of a flat-end cutter with radius r is represented in the cutter coordinate system (O_c, $\mathbf{x}_c$, $\mathbf{y}_c$, $\mathbf{z}_c$) by (see Fig. 3.2)

$$\mathrm{P}_c = \begin{bmatrix} r\sin\theta \\ -r\sin\theta \\ 0 \end{bmatrix}, \tag{3.17}$$

where θ denotes the angle from the negative $\mathbf{y}_c$ axis vector to the point P on the cutter's bottom edge (see Fig. 3.2).

For the flat-end tool, the center of the cutter coordinate system, O_c, is the coordinate of the cutter location point. Calculation of the CL point from CC data for flat-end cutter is straightforward. Initially, the CL point is represented in the local coordinate system as

$$\mathrm{CL} = \begin{bmatrix} -r \\ 0 \\ 0 \end{bmatrix}. \tag{3.18}$$

Then the tool is inclined by λ around the $\mathbf{y}_l$-axis. The CL point is now expressed as

[1] The tool cutting direction is the direction from the current CC point to the next CC point.

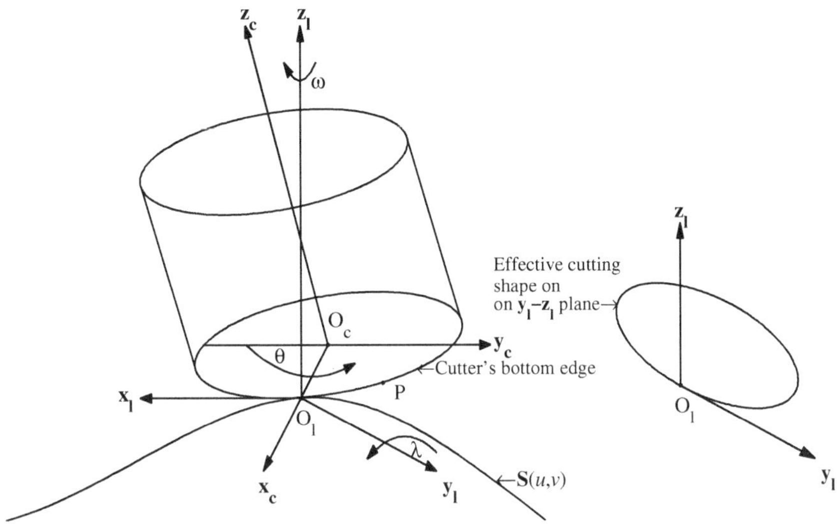

Fig. 3.2. Geometric analysis of the cutting operations

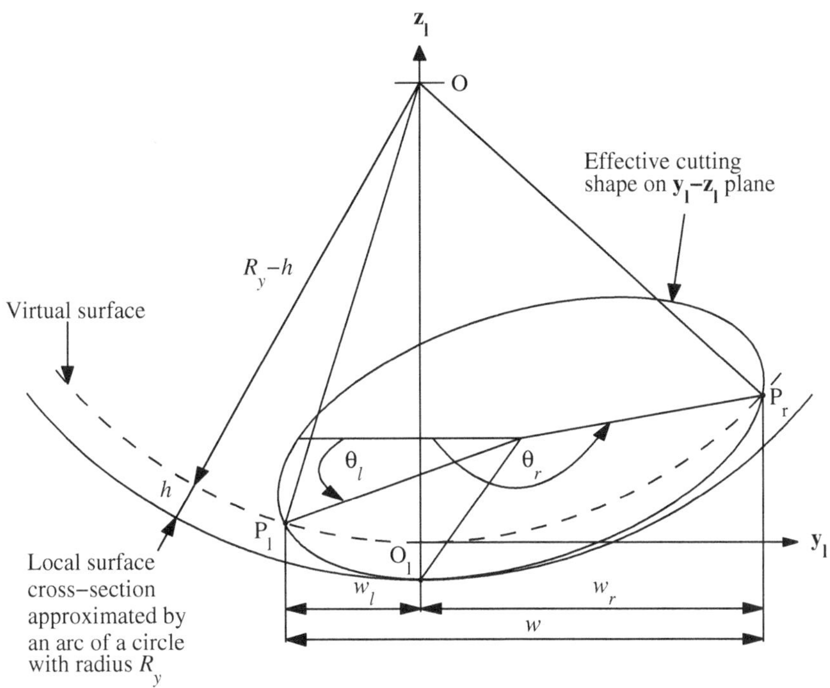

Fig. 3.3. Machining strip width estimation

$$\mathrm{CL} = \begin{bmatrix} \cos\lambda & 0 & \sin\lambda \\ 0 & 1 & 0 \\ -\sin\lambda & 0 & \cos\lambda \end{bmatrix} \begin{bmatrix} -r \\ 0 \\ 0 \end{bmatrix} = \begin{bmatrix} -r\cos\lambda \\ 0 \\ r\sin\lambda \end{bmatrix}. \tag{3.19}$$

Finally, tilting the tool by ω around $\mathbf{z}_l$ yields

$$\mathrm{CL} = \begin{bmatrix} \cos\omega & \sin\omega & 0 \\ -\sin\omega & \cos\omega & 0 \\ 0 & 0 & 1 \end{bmatrix} \begin{bmatrix} -r\cos\lambda \\ 0 \\ r\sin\lambda \end{bmatrix} = \begin{bmatrix} -r\cos\lambda\cos\omega \\ r\cos\lambda\sin\omega \\ r\sin\lambda \end{bmatrix}. \tag{3.20}$$

CL point is then expressed in workpiece coordinate system as

$$\mathrm{CL} = \mathrm{CC} + (-r\cos\lambda\cos\omega)\mathbf{x}_l + (r\cos\lambda\sin\omega)\mathbf{y}_l + (r\sin\lambda)\mathbf{z}_l. \tag{3.21}$$

Similarly, point P on the bottom edge of a flat-end cutter is represented in the local coordinate system (O_l, $\mathbf{x}_l$, $\mathbf{y}_l$, $\mathbf{z}_l$) by

$$\begin{aligned} \mathrm{P} &= \begin{bmatrix} x_l \\ y_l \\ z_l \end{bmatrix} \\ &= \begin{bmatrix} \cos\omega & \sin\omega & 0 \\ -\sin\omega & \cos\omega & 0 \\ 0 & 0 & 1 \end{bmatrix} \begin{bmatrix} \cos\lambda & 0 & \sin\lambda \\ 0 & 1 & 0 \\ -\sin\lambda & 0 & \cos\lambda \end{bmatrix} (\mathrm{O}_c + \mathrm{P}_c), \\ &= \begin{bmatrix} \cos\omega & \sin\omega & 0 \\ -\sin\omega & \cos\omega & 0 \\ 0 & 0 & 1 \end{bmatrix} \begin{bmatrix} \cos\lambda & 0 & \sin\lambda \\ 0 & 1 & 0 \\ -\sin\lambda & 0 & \cos\lambda \end{bmatrix} \left(\begin{bmatrix} -r \\ 0 \\ 0 \end{bmatrix} + \begin{bmatrix} r\sin\theta \\ -r\sin\theta \\ 0 \end{bmatrix} \right), \\ &= \begin{bmatrix} -r(\cos\lambda\cos\omega(1-\sin\theta) + \sin\omega\cos\theta) \\ r(\cos\lambda\sin\omega(1-\sin\theta) - \cos\omega\cos\theta) \\ r\sin\lambda(1-\sin\theta) \end{bmatrix}. \end{aligned} \tag{3.22}$$

To evaluate the machining strip width, the surface cross-section perpendicular to the tool cutting direction $\mathbf{x}_l$ is approximated by a circular arc, for which the radius R_y is equal to the radius of normal curvature[2] of the surface in the $\mathbf{y}_l$ direction as shown in Fig. 3.3. Given the maximum scallop height h, the maximum machined surface error is represented by a virtual circular arc of radius $R_y - h$ concentric with the first circular arc as shown in Fig. 3.3. The machining strip width is then obtained by locating the intersections of the effective cutting shape with this virtual circular arc.

The equation of the effective cutting shape is found by projecting (3.22) onto the $\mathbf{y}_l$-$\mathbf{z}_l$ plane, i.e., by setting x_l component of point P to zero. The equation of the virtual circular arc with radius $R_y - h$ as shown in Fig. 3.3 is given by

$$y_l^2 + (z_l - R_y)^2 = (R_y - h)^2. \tag{3.23}$$

---

[2] The radius of normal curvature is positive for concave surface and negative for convex surface.

Substituting (3.22) into (3.23) yields

$$c_1 \sin^2\theta + c_2 \sin\theta + c_3 \sin 2\theta - 2c_3 \cos\theta + c_4 = 0, \tag{3.24}$$

where

$$c_1 = r^2(cos^2\lambda \sin^2\omega + \sin^2\lambda - \cos^2\omega),$$
$$c_2 = -2r^2(cos^2\lambda \sin^2\omega + \sin^2\lambda) + 2rR_y \sin\lambda,$$
$$c_3 = \frac{r^2 \cos\lambda \sin 2\omega}{2},$$
$$c_4 = r^2 \sin^2\lambda \cos^2\omega - 2rR_y \sin\lambda + r^2 + 2hR_y - h^2.$$

The values of θ_l and θ_r corresponding to the left and right intersection points P_l and P_r (see Fig. 3.3) are found by solving (3.24).

When $\omega = 0$, (3.24) has a closed form solution given by

$$\begin{aligned} \theta_l &= \sin^{-1}\left(\frac{-c_2 - \sqrt{c_2^2 - 4c_1c_4}}{2c_1}\right), \ 0 \le \theta_l \le \frac{\pi}{2}, \\ \theta_r &= \pi - \theta_l. \end{aligned} \tag{3.25}$$

We are interested in the solutions θ_l and θ_r close to the CC point, i.e., the intersection points on the lower edge of an ellipse. When the inclination angle λ is small or the maximum scallop height h is large, (3.24) and (3.25) might exhibit no solutions or the intersection points lie on the upper edge of an ellipse. In such case, θ_l and θ_r are set to θ_{P_l} and θ_{P_r} corresponding to the left-most point and the right-most point on the effective cutting shape, respectively. The left-most and right-most points on the effective cutting shape are found by solving $dy_l(\theta)/d\theta = 0$. Differentiating y_l in (3.22) yields

$$r \cos\lambda \sin\omega \cos\theta - r \cos\omega \sin\theta = 0. \tag{3.26}$$

Therefore,

$$\begin{aligned} \theta_{\mathrm{P}_l} &= \tan^{-1}(\cos\lambda \tan\omega), \\ \theta_{\mathrm{P}_r} &= \tan^{-1}(\cos\lambda \tan\omega) + \pi. \end{aligned} \tag{3.27}$$

The left and the right machining strip widths w_l and w_r at the current CC point O_l are then expressed as

$$\begin{aligned} w_l &= |r(\cos\omega \cos\theta_l - \cos\lambda \sin\omega(1 - \sin\theta_l))|, \\ w_r &= |r(\cos\omega \cos\theta_r - \cos\lambda \sin\omega(1 - \sin\theta_r))|. \end{aligned} \tag{3.28}$$

The entire machining strip width is then defined by

$$w = w_l + w_r. \tag{3.29}$$

The machining strip width estimation method presented above is similar to the method used by Lee and Ji [11] but with two modifications. First, the local surface cross-section at the CC point is approximated by a circular arc instead of the quadratic equation (see Fig. 3.4(a)) for consistency with the subsequent calculation for optimal tool orientation and gouging avoidance presented in Sect. 3.3. Second, by approximating the surface cross-section by circular arc, the offset curve corresponding to the maximum allowable machined surface error h, can be easily computed by using a virtual circular arc of radius $R-h$ as shown in Fig. 3.3. In contrast, the machined surface error or scallop height, h, is underestimated when using the method of Lee and Ji where the offset curve is computed by shifting the approximated local surface cross-section in the normal direction $\mathbf{z}_l$ by h as shown in Fig. 3.4(a). The difference $\Delta h = h - \hat{h}$ is, however, negligible for small values of h (see Fig. 3.4(b)).

Machining strip width estimation for ball-end cutter

The calculation of the machining strip width of a ball-end cutter can be considered as a special case of the method presented above when $\lambda = \pi/2$ and $\omega = 0$. In this case (3.24) becomes

$$(-2r^2 + 2rR_y)\sin\theta + (2r^2 - 2rR_y) + 2hR_y - h^2 = 0, \tag{3.30}$$

$$\sin\theta = 1 - \frac{2hR_y - h^2}{2rR_y - 2r^2}. \tag{3.31}$$

Using the trigonometric identity $\cos^2\alpha + \sin^2\alpha = 1$, the machine strip width of (3.28) becomes

$$\begin{aligned} w_l = w_r &= r\cos\theta, \\ &= r\sqrt{1 - \sin^2\theta}. \end{aligned} \tag{3.32}$$

The entire machining strip width is then given by

$$w = w_l + w_r = 2r\sqrt{\frac{2hR_y - h^2}{rR_y - r^2} - \left(\frac{2hR_y - h^2}{2rR_y - 2r^2}\right)^2}. \tag{3.33}$$

Since in practice $|R_y| \gg h$, the terms consisting of h^2, h^3, and h^4 in (3.33) can be disregarded. The maching strip width is then approximated by

$$w = \sqrt{\frac{8hrR_y}{R_y - r}}, \tag{3.34}$$

which is the same approximation presented in [12]. Equation (3.34) is also used in [13] to calculate the machining strip width for a flat-end tool where the tool radius r is replaced by the effective cutter radius r_e (see Sect. 3.3).

The machining strip width calculation of the toroidal cutter can be found in [2].

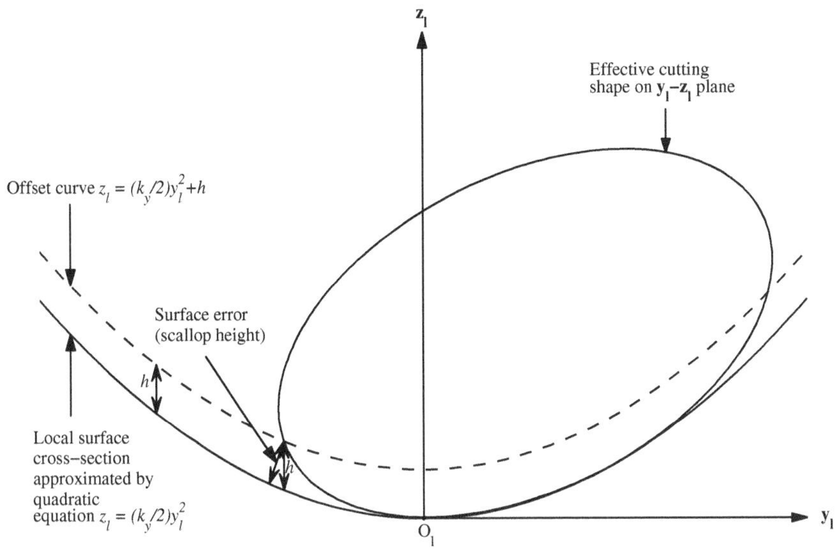

(a)

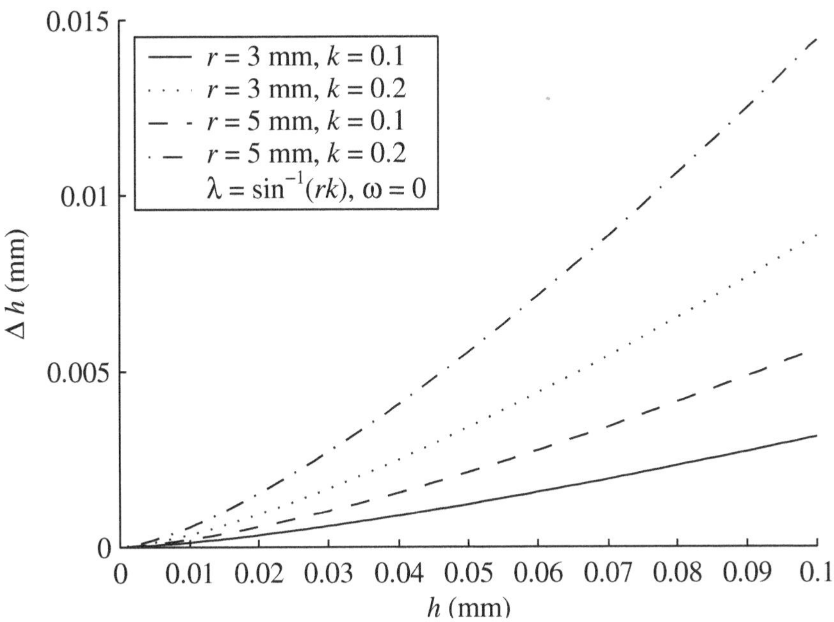

(b)

Fig. 3.4. (**a**) machining strip width estimation method of Lee and Ji, (**b**) plot of the difference between the maximum allowable error h, and the estimated machined surface error $\hat{h}$, with normal curvature k, tool radius r, and tool orientation $(\lambda, \omega) = (\sin^{-1}(rk), 0)$

3.3 Optimal Tool Orientation and Gouging Avoidance

In five-axis machining, the effective cutting shape is determined by the tool orientation (λ, ω). The effective cutter radius r_e of a flat-end cutter is given by [13]

$$r_e = ra^2 \left(\frac{1 + b^2}{a^2 + b^2} \right)^{3/2}, \tag{3.35}$$

where

$$a = \sin\lambda \cos\omega,$$

$$b = \tan\lambda \sin\omega,$$

which is simplified to

$$r_e = \frac{r\cos^2\omega}{\sin\lambda}. \tag{3.36}$$

To optimize the machining strip width, λ and ω are usually set so that r_e is best matched to the radius of curvature at the CC point. For convex or planar surfaces (see Sect. 3.1), the tool inclination angle λ is normally set to a small default angle or zero and the tilt angle ω is set to zero. If the surface is non-convex, a non-zero tool inclination angle λ is needed to avoid gouging.

Lee and Ji [11] presents a simple and fast algorithm suitable for being incorporated into the tool path generation strategies. The algorithm finds a minimum inclination angle to eliminate overcutting in both the $\mathbf{x}_l$-$\mathbf{z}_l$ and $\mathbf{y}_l$-$\mathbf{z}_l$ planes, as follows:

$$\begin{aligned} \lambda_{\text{min}} &= \max(\lambda_x, \lambda_y), \\ \lambda_x &= \sin^{-1}\left(\frac{r}{R_x}\right), \quad \text{if } (r \le R_x) \text{ and } (R_x > 0), \\ \lambda_y &= \sin^{-1}\left(\frac{r}{R_y}\right), \quad \text{if } (r \le R_y) \text{ and } (R_y > 0), \end{aligned} \tag{3.37}$$

where R_x and R_y are the radius of normal curvature in the $\mathbf{x}_l$ and $\mathbf{y}_l$ directions, respectively.

Unfortunately, the method is not applicable to the non-convex surfaces when R_x and R_y are both negative but the maximum principle curvature is positive. Clearly, in this case, the method outputs $\lambda_{\text{min}} = 0$. This "bug" often leads to local gouging. Lo [13] solves this problem by continuously checking for gouging in all directions. Furthermore, a modification of these techniques is presented in this book.

Consider a flat-end cutter as shown in Fig. 3.2. Gouging occurs when any point on the circle touches or goes inside the surface. Let G be a gouging point, which is another point on the circle that touches the surface (as shown in Fig. 3.5(a)). The line connecting the two points, O_l and G, forms a chord on the circle and has an angle of ϕ with respect to $\overline{\text{O}_l\text{O}_c}$ (see Fig. 3.5(a)). Let λ_ϕ be the tool inclination angle that corresponds to a specific value of ϕ

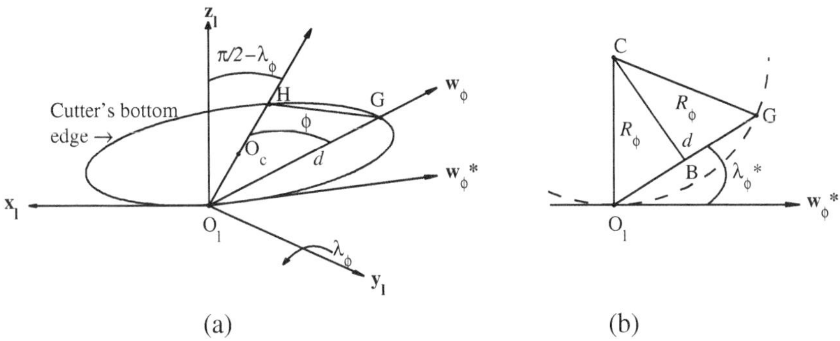

Fig. 3.5. Tool gouging

for a gouging point G, the minimum tool inclination angle to avoid gouging is formulated as the following optimization problem:

$$\lambda_{\min} = \max_{-\pi/2 \leq \phi \leq \pi/2} \lambda_\phi. \tag{3.38}$$

Lo [13] solved this problem numerically by utilizing the gradient method. However, it can be shown that the solution of (3.38) for a non-convex surface can be expressed in a closed-form formula.

Consider a gouging point G, the chord $\overline{\mathrm{O}_l G}$ has a length of d and an angle of ϕ with respect to $\overline{\mathrm{O}_l \mathrm{O}_c}$ (see Fig. 3.5(a)). Since O_lGH is a right-angled triangle having the angle O_lGH right, the length d can be calculated by

$$d = 2r\cos\phi. \tag{3.39}$$

Let $\mathbf{w}_\phi$ be a unit vector in the direction of $\overline{\mathrm{O}_l G}$ and $\mathbf{w}_\phi^*$ be a unit vector tangent to the surface which is obtained by projecting $\mathbf{w}_\phi$ onto the $\mathbf{x}_l$-$\mathbf{y}_l$ plane as shown in Fig. 3.5(a). Since O_l and G are both located on the surface, they can be connected by a circular arc whose radius R_ϕ is equal to the radius of curvature in the $\mathbf{w}_\phi^*$ direction (see Fig. 3.5(b)). Let λ_ϕ be the inclination angle of the tool that corresponds to a specific value of ϕ for a gouging point G, $\mathbf{w}_\phi$ can be obtained by rotating the unit vector $[-1, 0, 0]$ by ϕ around the $\mathbf{z}_l$-axis and then by λ_ϕ around the $\mathbf{y}_l$-axis. Thus, we have

$$\begin{aligned}\mathbf{w}_\phi &= \begin{bmatrix} \cos\lambda_\phi & 0 & \sin\lambda_\phi \\ 0 & 1 & 0 \\ -\sin\lambda_\phi & 0 & \cos\lambda_\phi \end{bmatrix} \begin{bmatrix} \cos\phi & \sin\phi & 0 \\ -\sin\phi & \cos\phi & 0 \\ 0 & 0 & 1 \end{bmatrix} \begin{bmatrix} -1 \\ 0 \\ 0 \end{bmatrix}, \\ &= \begin{bmatrix} -\cos\phi\cos\lambda_\phi \\ \sin\phi \\ \cos\phi\sin\lambda_\phi \end{bmatrix}. \end{aligned} \tag{3.40}$$

$\mathbf{w}_\phi^*$ is expressed as

$$\mathbf{w}_\phi^* = \begin{bmatrix} \dfrac{-\cos\phi\cos\lambda_\phi}{\sqrt{\cos^2\phi\cos^2\lambda_\phi + \sin^2\phi}} \\ \dfrac{\sin\phi}{\sqrt{\cos^2\phi\cos^2\lambda_\phi + \sin^2\phi}} \\ 0 \end{bmatrix}. \tag{3.41}$$

Let λ_ϕ^* be the angle by which $\overline{O_l G}$ is inclined with respect to the $\mathbf{x}_l$-$\mathbf{y}_l$ plane as shown in Fig. 3.5(b). From (3.40) and (3.41), we have

$$\cos\lambda_\phi^* = \mathbf{w}_\phi \cdot \mathbf{w}_\phi^* = \sqrt{\cos^2\phi\cos^2\lambda_\phi + \sin^2\phi}. \tag{3.42}$$

Since $\overline{CO_l}$ is perpendicular to the $\mathbf{x}_l$-$\mathbf{y}_l$ plane and $O_l BC$ is a right-angled triangle having the angle $O_l BC$ right, the angle $O_l CB$ is, therefore, equal to λ_ϕ^*. The length of $\overline{O_l G}$ is then given by

$$\begin{aligned} d &= 2R_\phi \sin\lambda_\phi^*, \\ &= 2R_\phi\sqrt{1-\cos^2\lambda_\phi^*}, \end{aligned} \tag{3.43}$$

where the value of $\sin\lambda_\phi^*$ is obtained using the trigonometric identity $\sin^2\alpha + \cos^2\alpha = 1$. Subsitituting (3.42) into (3.43) yields

$$\begin{aligned} d &= 2R_\phi\sqrt{1-\sin^2\phi - \cos^2\phi\cos^2\lambda_\phi}, \\ &= 2R_\phi\sqrt{\cos^2\phi - \cos^2\phi\cos^2\lambda_\phi}, \\ &= 2R_\phi\cos\phi\sqrt{1-\cos^2\lambda_\phi}, \\ &= 2R_\phi\cos\phi\sin\lambda_\phi. \end{aligned} \tag{3.44}$$

Equating (3.39) to (3.44), we have

$$\lambda_\phi = \sin^{-1}(r/R_\phi) = \sin^{-1}(rk_\phi). \tag{3.45}$$

The minimum inclination angle λ for gouging avoidance is then expressed as

$$\lambda_{\text{min}} = \max_{-\pi/2\le\phi\le\pi/2} \lambda_\phi = \max_{-\pi/2\le\phi\le\pi/2} \sin^{-1}(rk_\phi) = \sin^{-1}(rk_{\text{max}}), \tag{3.46}$$

where k_{max} is the maximum surface curvature at the CC point.

In the case when gouging can not be eliminated by inclining the tool alone or the inclination angle λ exceeds the limit of the machine, searching for the tilt angle ω is needed or a smaller tool size must be used [7]. An algorithm for setting the tilt angle ω can be found in [11].

Remark 3.1. The surface cross-section is approximated by a circular arc in the $\mathbf{y}_l$-$\mathbf{z}_l$ plane corresponding to an osculating circle having the radius R_y. The osculating circle represents the best approximating curve having the constant curvature. The approach has been used in many practical applications (see, for instance, [12, 13]). However, the approximation is accurate only in the vicinity of the CC point. Therefore, the estimate of the machining strip is accurate when the machining strip itself is sufficiently small. The applicability of the approximation can be verified by comparing the approximating arcs with the actual curves on the surface in an appropriate norm. If the accuracy is insufficient then a high order approximation must be used.

Remark 3.2. The accuracy of the above model based on a single contact point may be insufficient. In this case multipoint strategies [20–22] could also be applied to further enhance the accuracy of the tool positioning.

3.4 Kinematics Error

Besides controlling the size of the scallop left between two successive tool paths, the forward step (kinematics) error due to nonlinearity of the machine kinematics needs to be minimized as well. The kinematics error between two CC points is defined as the difference between the desired and the actual trajectory of the tool tip. The mathematical representation of the tool tip trajectory can be derived as follows:

Let $\mathfrak{K}$ be the transformation such that $\forall\, W, M, \mathfrak{R}$, $\mathfrak{K}^{-1}(\mathfrak{K}(\mathfrak{R}, M)) = M$ and $\mathfrak{K}(\mathfrak{K}^{-1}(\mathfrak{R}, W)) = W$. Let $\Pi_p \equiv (M_p, \mathfrak{R}_p)$, $\Pi_{p+1} \equiv (M_{p+1}, \mathfrak{R}_{p+1})$ be two successive coordinates of the tool path in $\mathbb{R}^5$. M_p and W_p denote spatial positions of the tool tip in machine and workpiece coordinate systems, respectively, and $\mathfrak{R}_p$ the corresponding pair of rotation angles.

Let us invoke the inverse kinematics to transform workpiece coordinates W_p into machine coordinates $M_p \equiv (x_p, y_p, z_p)$ as follows: $M_p \equiv \mathfrak{K}^{-1}(\mathfrak{R}_p, W_p)$. The rotation angles $\mathfrak{R} \equiv \mathfrak{R}(t) = (a(t), b(t))$ and the machine coordinates of the tool tip $M \equiv M(t)$ are assumed to change linearly between the prescribed points, namely, $M(t) = tM_{p+1} + (1-t)M_p$, $\mathfrak{R}(t) = t\mathfrak{R}_{p+1} + (1-t)\mathfrak{R}_p$, where t is a fictitious time coordinate ($0 \leq t \leq 1$). Transforming machine coordinates M back to workpiece coordinates W for every t yields

$$
\begin{aligned}
W_{p,p+1}(t) &= \mathfrak{K}(\mathfrak{R}(t), M(t)), \\
&= \mathfrak{K}(\underline{t\mathfrak{R}_{p+1} + (1-t)\mathfrak{R}_p}, \underline{tM_{p+1} + (1-t)M_p}).
\end{aligned} \tag{3.47}
$$

Now the machine coordinates M_p and M_{p+1} must be eliminated so that the resulting trajectory depends only on the workpiece coordinates and orientation of the tool. M_p, M_{p+1} are eliminated by using the inverse transformation $M_p = \mathfrak{K}^{-1}(\mathfrak{R}_p, W_p)$. Substituting M_p, M_{p+1} into (3.47) yields

$$
\begin{aligned}
W_{p,p+1}(t) = \mathfrak{K}(&\underline{t\mathfrak{R}_{p+1} + (1-t)\mathfrak{R}_p}, \\
&\underline{t\mathfrak{K}^{-1}(\mathfrak{R}_{p+1}, W_{p+1}) + (1-t)\mathfrak{K}^{-1}(\mathfrak{R}_p, W_p)}).
\end{aligned} \tag{3.48}
$$

Let $W^D_{p,p+1}(t) \equiv (x^D_{p,p+1}(t), y^D_{p,p+1}(t), z^D_{p,p+1}(t)) \in S(u,v)$ be a curve between W_p and W_{p+1} extracted from the surface in such a way that it represents the desired tool trajectory. The kinematics error is represented as a maximum deviation between $W^D_{p,p+1}(t)$ and $W_{p,p+1}(t) \equiv (x_{p,p+1}(t), y_{p,p+1}(t), z_{p,p+1}(t))$ (see Fig. 3.6), namely,

$$\begin{aligned}\epsilon_{p,p+1} &= \max_{0\le t\le 1} |W^D_{p,p+1}(t) - W_{p,p+1}(t)|, \\ &= \max_{0\le t\le 1} \Big[\big(x^D_{p,p+1}(t) - x_{p,p+1}(t)\big)^2 + \big(y^D_{p,p+1}(t) - y_{p,p+1}(t)\big)^2 \\ &\qquad + \big(z^D_{p,p+1}(t) - z_{p,p+1}(t)\big)^2\Big]^{\frac{1}{2}}.\end{aligned} \tag{3.49}$$

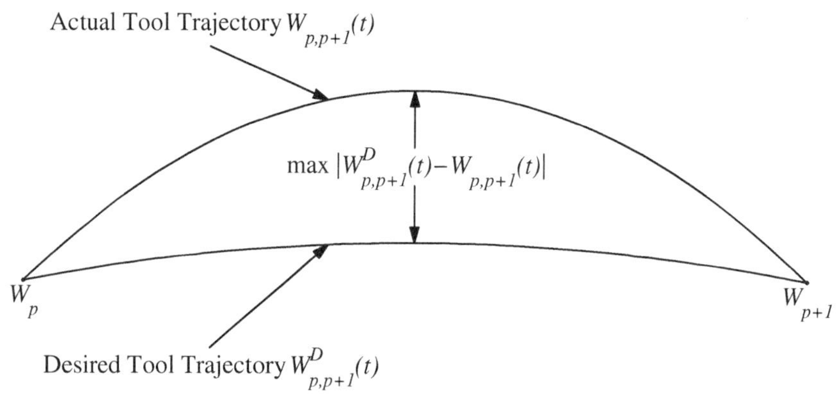

Fig. 3.6. Kinematics error between two cutter contact points W_p and W_{p+1}

The kinematics error depends not only on the characteristics of the surface versus the tool orientation but on the previous rotations as well [15]. Consider the machine kinematics given in (2.2) and (2.5) for the 2-0 machine. If the size of angular jump of rotation angle a between any two points is larger than π, then the subsequence value of a should be adjusted in order to minimize the difference eliminating unexpected motion and collision due to sharp variations of the rotation angle a. As an example, angle adjustment for the 2-0 machine can be performed without effecting the desired tool orientation by using one of the following rules (see Fig. 3.7 and [15]).

$$\begin{aligned}(1)\ a_{\text{new}} &= a_{\text{old}} \pm \pi, \\ b_{\text{new}} &= -\pi - b_{\text{old}}, \\ (2)\ a_{\text{new}} &= a_{\text{old}} \pm 2\pi, \\ b_{\text{new}} &= b_{\text{old}},\end{aligned} \tag{3.50}$$

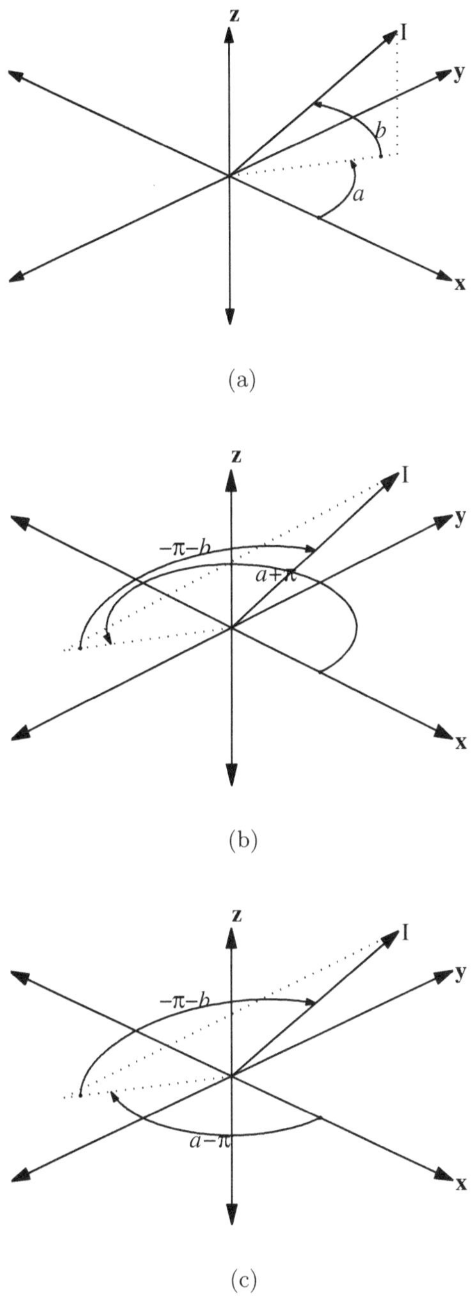

Fig. 3.7. Angle adjustment for the 2-0 machine

The above rules can be interpreted as follows. For the 2-0 machine, the tool orientation relative to the workpiece can be described by two angles, a and b. The tool which is initially aligned with the **x**-axis is first rotated around **y**-axis by b and then around **z**-axis by a (see Fig. 3.7(a)). The same tool orientation can also be achieved by first rotating the tool around **y**-axis by $-\pi - b$ and then around **z**-axis by $a + \pi$ (see Fig. 3.7(b)). Likewise, the same tool orientation can also be achieved by first rotating the tool around **y**-axis by $-\pi - b$ and then around **z**-axis by $a - \pi$ (see Fig. 3.7(c)).

Figure 3.8 demonstrates the minimization of kinematics error by the angle adjustment rules. Consider two CC points, W_1 and W_2, and the associated tool orientations I_1 and I_2 given by

$$\begin{aligned} W_1 &= [-50.0, -50.0, -40.0], \\ I_1 &= [0, -0.625, 0.781], \\ W_2 &= [-50.0, 10.0, -20.8], \\ I_2 &= [0, 0.158, 0.987]. \end{aligned} \tag{3.51}$$

The machine rotation angles are calculated using (2.5).

$$\begin{aligned} (a_1, b_1) &= (4.7122, -0.896), \\ (a_2, b_2) &= (1.571, -1.412). \end{aligned} \tag{3.52}$$

The tool trajectory is shown in Fig. 3.8(a). To reduce the error, a_2 and b_2 are adjusted as follows:

$$\begin{aligned} a_2^{\text{new}} &= a_2 + \pi = 4.7126, \\ b_2^{\text{new}} &= -\pi - b_2 = -1.7296. \end{aligned} \tag{3.53}$$

The tool trajectory after angle adjustment is shown in Fig. 3.8(b). Angle adjustment can be simultaneously applied to a set of successive points on the tool path (see Chap. 6). However, if the kinematics error between the two points still exceeds a prescribed tolerance, a new point is inserted between the two (see Chap. 6).

3.5 Tool Path Generation

Sections 3.1 to 3.4 provide necessary information needed for tool path generation. There exists a large number of tool path generation techniques presented in the literature (see surveys by Dragomatz and Mann [3] and Sarma [17]). Tool path generation techniques can be classified into two main groups: isoparametric tool path (see, for examples, [1, 4, 19]) and non-isoparametric tool paths (see, for examples, [5, 6, 10, 12, 14]).

The isoparametric method is popular for tool path generation due to ease of computation. However, the machining efficiency and surface smoothness

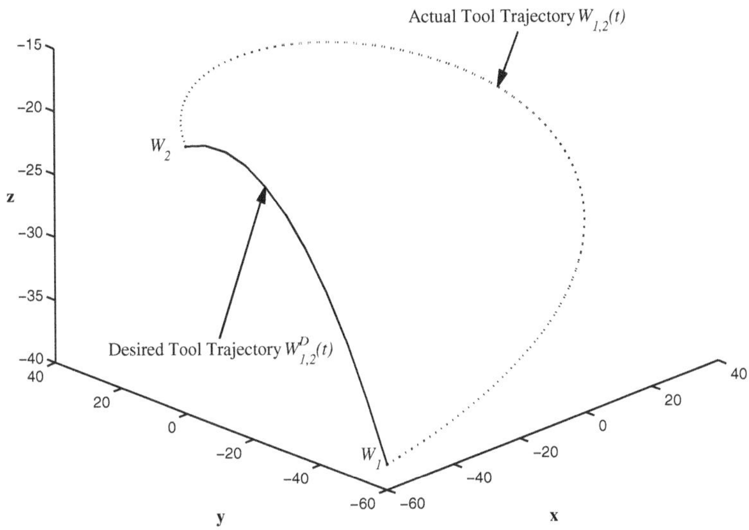

(a)

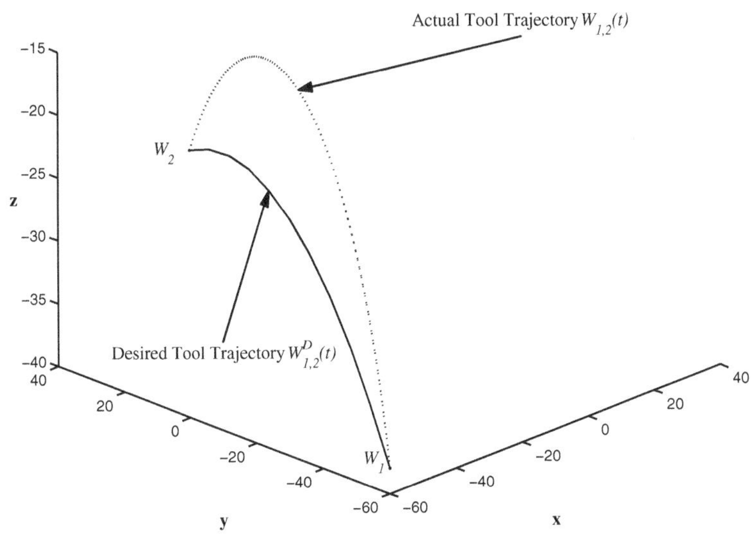

(b)

Fig. 3.8. Kinematics error reduction by angle adjustment can be quite poor due to the redundant machining and an unpredictable scallop height, especially when the parametric increment is too small or large. Isoparametric tool paths are generated by calculating the smallest tool path interval and using it as a constant offset to the next tool path. The tool path interval depends on the local surface shape, cutter shape and size, and the allowable scallop height (see Sect. 3.2). Besides, the machining strip widths of the two adjacent cutter paths (Fig. 3.9) have to overlap to ensure that the machined surface error (scallop height) is within the tolerance. The incremental values Δu and Δv in the parametric domain corresponding to the machining strip width w can be found by solving the following equation.

$$\Delta u S_u + \Delta v S_v = w \mathbf{y}_l. \tag{3.54}$$

Example of isoparametric tool path is shown in Fig. 3.10.

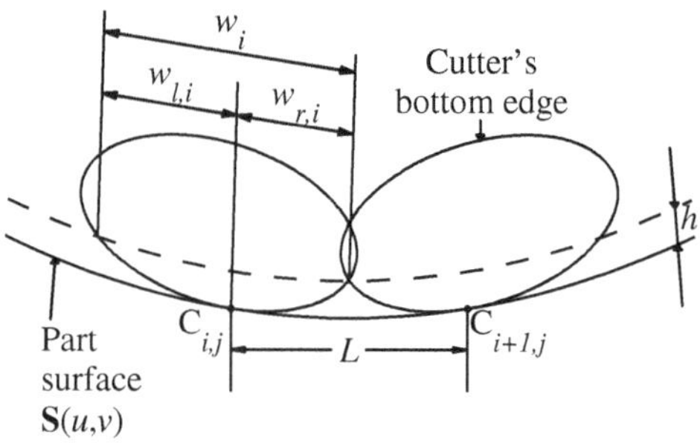

Fig. 3.9. Overlapping of machining strips on adjacent tool paths

Optimization of the tool path of a five-axis NC machine presents a considerable challenge. Recent papers have displayed a number of sophisticated methods to optimize the conventional isoparametric tool path. Besides, a variety of methods is available to generate unconventional patterns, for instance: the neural network approach [18], the Voronoi diagram technique [8], the monotone chain method [16], the distance map method [9], etc. The next two chapters present two techniques for tool path generation and optimization, namely, the adaptive space-filling curves and the grid generation method. Each technique can be used to optimize the conventional isoparametric tool path. Furthermore, when combining the two techniques together, an efficient tool path can be generated for surfaces with complicated boundaries, cuts off, pockets, islands, etc.

References

[1] Bobrow, J. E. 1985. NC machine tool path generation from CSG part representations. *Computer Aided Design*, 17(2):69–76.

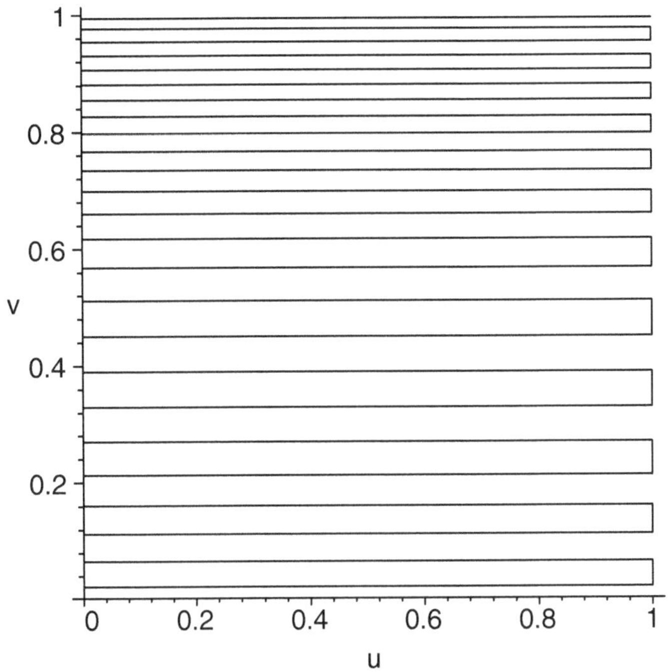

(a)

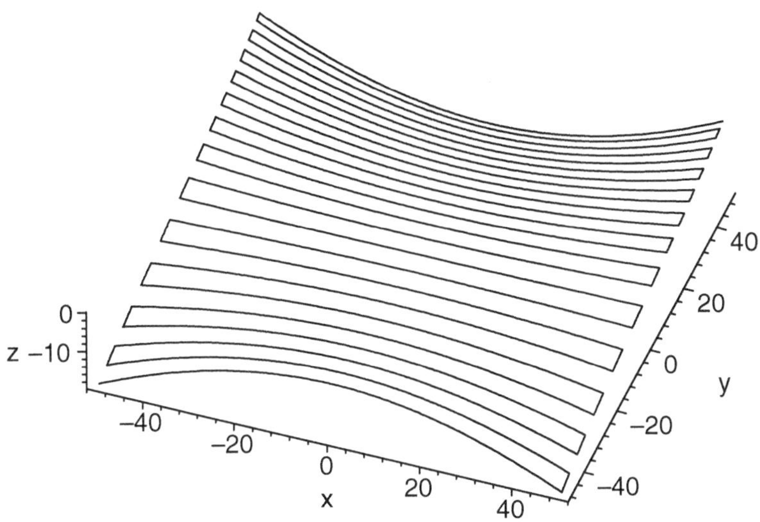

(b)

Fig. 3.10. Example of isoparametric tool path

[2] Chiou, C.-J. and Lee, Y.-S. 2002. A machining potential field approach to tool path generation for multi-axis sculptured surface machining. *Computer-Aided Design*, 34(5):357–371.

[3] Dragomatz, D. and Mann, S. 1997. A classified bibliography of literature on NC milling path generation. *Computer-Aided Design*, 29(3):239–247.

[4] Elber, G. and Cohen, E. 1994. Toolpath generation for freeform surface models. *Computer-Aided Design*, 26(6):490–496.

[5] Griffiths, J. G. 1994. Toolpath based on Hilbert's curve. *Computer-Aided Design*, 26(11):839–844.

[6] Hatna, A. and Grieve, B. 2000. Cartesian machining versus parametric machining: a comparative study. *International Journal of Production Research*, 38(13):3043–3065.

[7] Jensen, C. G., Red, W. E., and Pi, J. 2002. Tool selection for five-axis curvature matched machining. *Computer-Aided Design*, 34(3):251–266.

[8] Jeong, J. and Kim, K. 1999a. Generating tool paths for free-form pocket machining using z-buffer-based Voronoi diagrams. *International Journal of Advanced Manufacturing Technology*, 15(3):182–187.

[9] Jeong, J. and Kim, K. 1999b. Generation of tool paths for machining free-form pockets with islands using distance maps. *International Journal of Advanced Manufacturing Technology*, 15(5):311–316.

[10] Lee, Y.-S. 1998. Non-isoparametric tool path planning by machining strip evaluation for 5-axis sculptured surface machining. *Computer-Aided Design*, 30(7):559–570.

[11] Lee, Y.-S. and Ji, H. 1997. Surface interrogation and machining strip evaluation for 5-axis CNC die and mold machining. *International Journal of Production Research*, 35(1):225–252.

[12] Lin, R.-S. and Koren, Y. 1996. Efficient tool-path planning for machining free-form surfaces. *ASME Journal of Engineering for Industry*, 118(1):20–28.

[13] Lo, C. C. 1999. Efficient cutter-path planning for five-axis surface machining with a flat-end cutter. *Computer-Aided Design*, 31(9):557–566.

[14] Makhanov, S. S., Batanov, D., Bohez, E., Sonthipaumpoon, K., Anotaipaiboon, W., and Tabucanon, M. 2002. On the tool-path optimization of a milling robot. *Computers & Industrial Engineering*, 43(3):455–472.

[15] Munlin, M.-A., Makhanov, S. S., and Bohez, E. L. J. 2004. Optimization of rotations of a five-axis milling machine near stationary points. *Computer-Aided Design*, 36(12):1117–1128.

[16] Park, S. C. and Choi, B. K. 2000. Tool-path planning for direction-parallel area milling. *Computer-Aided Design*, 32(1):17–25.

[17] Sarma, R. 2000. An assessment of geometric methods in trajectory synthesis for shape-creating manufacturing operations. *Journal of Manufacturing Systems*, 19(1):59–72.

[18] Suh, S. H. and Shin, Y. S. 1996. Neural network modeling for tool path planning of rough cut in complex pocket milling. *Journal of Manufacturing Systems*, 15(5):295–304.

[19] Wang, Y. and Tang, X. 1999. Five-axis NC machining of sculptured surfaces. *International Journal of Advanced Manufacturing Technology*, 15(1):7–14.

[20] Warkentin, A., Ismail, F., and Bedi, S. 1998. Intersection approach to multi-point machining of sculptured surfaces. *Computer Aided Geometric Design*, 15(6):567–584.

[21] Warkentin, A., Ismail, F., and Bedi, S. 2000. Multi-point tool positioning strategy for 5-axis machining of sculptured surfaces. *Computer Aided Geometric Design*, 17(1):83–100.

[22] Yoon, J.-H. 1997. Tool tip gouging avoidance and optimal tool positioning for 5-axis sculptured surface machining. *International Journal of Production Research*, 41(10):2125–2142.

4

Space-Filling Curve Tool Paths

4.1 History of Space-Filling Curves and Their Applications

An N-dimensional space-filling curve (SFC) is a continuous, surjective[1] (onto) function from the unit interval $[0, 1]$ to the N-dimensional unit hypercube $[0, 1]^N$. In particular, a 2-dimensional space-filling curve is a continuous curve that passes through every point of the unit square $[0, 1]^2$.

The history of space-filling curves started in 1878 when George Cantor (1845–1918) demonstrated that any two smooth manifolds of arbitrarily finite dimensions have the same cardinality. Cantor's finding implies that the unit line segment $[0, 1]$ can be mapped bijectively[2] onto the unit square $[0, 1]^2$. In 1879, Eugen Netto (1848–1919), however, demonstrated that such mapping is necessary discontinuous and cannot be called a curve. If the condition of bijectivity were dropped, Giuseppe Peano (1858–1932) found a continuous map from the interval onto the square in 1890. This was the first example of a space-filling curve (see Fig. 4.1). Furthers examples were introduced by D. Hilbert (in 1891, see Fig. 4.2), E.H. Moore (in 1900), H. Lebesgue (in 1904), W. Sierpiński (in 1912), G. Pólya (in 1913), etc [23].

SFCs are encountered in different fields of computer science, especially where it is important to linearize multidimensional data. Examples of multi-dimensional data are matrices, images, tables and computational grids resulting from the discretization of partial differential equations (PDEs). Typical applications of SFCs are data indexing [14, 20], data storing and retrieving [25], image processing [26, 28], image scanning and coding [5, 7, 27], mesh partitioning and reordering [24], etc.

[1] A function f from a domain X to a codomain Y is said to be surjective if its values span its whole codomain; that is, for every y in Y, there is at least one x in X such that $f(x) = y$.

[2] A function f from a domain X to a codomain Y is said to be bijective if for every y in Y there is exactly one x in X such that $f(x) = y$.

Fig. 4.1. 3 iterations of the Peano's space-filling curve

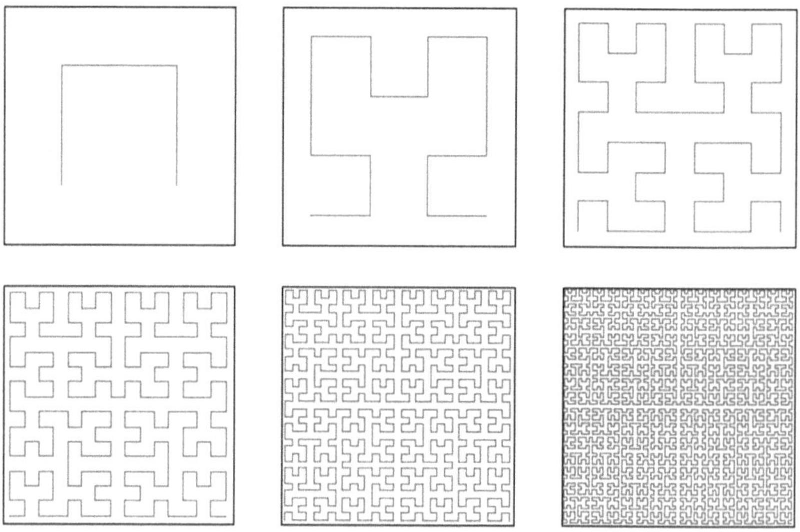

Fig. 4.2. 6 iterations of the Hilbert's space-filling curve

With the variety of space-filling curves and the wide spread of multi-dimensional applications, the selection of the appropriate space-filling curve for a certain application is not a trivial task. According to the classification in [1], space-filling curves are classified into two categories: recursive and non-recursive space-filling curves. Examples of recursive SFCs are the Peano's curve (Fig. 4.1) and the Hilbert's curve (Fig. 4.2). Most existing applications of SFCs employ recursive SFCs. The recursive behavior of SFCs allows for the linearization of recursive hierarchical data structures.

One of the most favorable properties of SFCs is their locality and the fact that the linearization is easily computable. Locality means that an SFC never leaves a region at any level of refinement before traversing all points of that region. Thus neighboring data in a multidimensional space remain neighboring after linearization.

In this chapter, we present the use of space-filling curves as tool paths. Given a set of CC data distributed on the surface, a space-filling curve tool path can be thought of as a curve that visits each CC point on the surface once and only once. We also consider the use of non-recursive SFC to allow the curve to travel in the optimal cutting directions.

4.2 Tool Path Optimization

Programming a five-axis NC machine requires consideration of a number of important issues, such as optimizing tool paths, avoiding interferences, and improving the cutting efficiency. Tool path optimization could be performed with regard to the cutting time, scallop heights, length of the tool path, width of the machining strip, volume of the removed material, etc. A complete optimization scheme involves specifying 1) a model of cutting operations, 2) topologies of the prescribed tool path patterns, 3) a set of constraints, and 4) an optimization procedure (see, for instance, [18]).

Let $S \equiv S(u,v)$ be the required surface, $\mathfrak{M}$ be a set of parameters related to the workpiece setup and the machine configuration, and $\mathfrak{T}$ be the parameters of the tool such as the tool length, diameter, shape. A model of the cutting operations, taking as input S, $\mathfrak{T}$, $\mathfrak{M}$ and Π produces a machining result, the output surface $T \equiv T(u,v)$. The general optimization problem is then formulated by

$$\min_{\Pi,\mathfrak{T},\mathfrak{M}} \epsilon, \tag{4.1}$$

where $\epsilon = \|S - T\|$ represents the error in an appropriate norm. The optimization is performed with respect to the following constraints:

1. Scallop height: the scallop between the successive tool tracks must not exceed a prescribed tolerance [16, 18].
2. Local accessibility: excess material must not be removed when the tool comes in contact with the desired surface. The excess removal commonly occurs due to *curvature interference* and *surface interference* [10, 11, 21].
3. Global accessibility: the tool must not come into contact with either machine parts or unwanted parts of the desired surface [13]. This means collision detection must be incorporated into the optimization procedure.

The set of independent variables comprises the tool spatial positions and orientations, rotations of the machine parts, the shape and the size of the tool. Usually, the tool visits the prescribed positions following the zigzag or spiral patterns. However, the optimization could also utilize complicated patterns adapted in such a way that a certain criteria is minimized or at least decreased.

We present the use of space-filling curves as tool paths for five-axis NC machining of sculptured surfaces. Application of the SFC to NC tool path generation has been first reported in [8] and [6]. Griffiths [8] proposed the use of the Hilbert curve as a tool path while Cox et al. [6] used various forms of space-filling curves such as the Moore curve. Both tool path generation methods have been developed for three-axis NC machining with the ball-end cutter. However, the space-filling curves have not been particularly popular in the five-axis machining community due to large inaccuracies produced by sharp angular turns which characterize the standard SFC patterns.

Nonetheless, tool path generation based on the SFC has a number of attractive features such as the possibility to locally adapt the curve (without changing the global structure) in such a way that the cutting device travels along the optimal direction. In addition, the entire surface is cut in one path eliminating the need of tool retractions.

The use of adaptive SFC as tool paths is presented in this book to overcome the above mentioned drawbacks while keeping the above mentioned advantages. The adaptive SFC is a better candidate for creating tool path than the Hilbert's curve. First, the Hilbert's curve is only applicable to a grid 2^n by 2^n while the adaptive SFC can be used for any rectangular grid as well as block structured grid (the generation of SFC tool paths on block structured grid is presented in Chap. 5). Second, as opposed to the Hilbert's curve having the ratio of three-quarters turn per point [12], the adaptive SFC turns only when necessary, in other words, only when the optimal cutting direction changes. Since the sharp turns usually produce large kinematics errors [19], it is anticipated that the adaptive SFC performs better than a conventional SFC (see also the forthcoming discussion in Sect. 4.3).

The SFC tool path generation method requires three steps: (1) the grid construction, (2) the generation of the adaptive space-filling curve, and (3) the tool path correction. The simplest rectangular grid can be constructed by overlaying two zigzag tool paths generated by the traditional isoparametric scheme. The grid is then filled by an SFC generated by the Hamiltonian path algorithm. Finally, the tool path and the orientation of the tool are adjusted at the turns to eliminate sharp variations in the orientation of the cutting device which result in large kinematics errors.

Consider the following optimization with regard to the length of the tool path,

$$\min_{\Pi} L, \quad \text{subject to } h < h_{\max}, \ \epsilon < \epsilon_{\max}, \tag{4.2}$$

where Π is the tool path represented by a structured set of the positions and orientations (the so-called CL data, see Sect. 2.1), h is the scallop height and $h_{\max}$ is the maximum allowed scallop height. If the machining strips do not overlap, the remaining areas are considered as scallops as well. Therefore, the first constraint controls the scallops and provides that the machining strips cover the entire surface. The second constraint requires that the difference ϵ between the actual and the desired trajectory does not exceed a prescribed tolerance $\epsilon_{\max}$. The above two conditions are nothing else than an approximation of a more general requirement that the machined surface does not deviate from the actual surface by more than $(h_{\max}, \epsilon_{\max})$. The minimization of L

is performed in such a way that Π maximizes the machining strip width. In turn, increasing the machining strip width combined with the SFC strategies makes it possible to cut the surface by traveling along a shorter path.

The scallop height constraint is satisfied by constructing two overlaying isoparametric zigzag patterns. The patterns satisfy the scallop height constraints given the surface curvature, a shape and a size of the cutter. The second constraint is satisfied by removing sharp angular turns producing large error and adding additional cutter-contact points, if necessary.

4.3 Tool Path Generation using Adaptive Space-filling Curves

The most popular SFC is the recursive Hilbert's curve [12] considered for numerous applications including the tool path planning [8]. Hilbert's curve is particularly appealing in tool path planning as its refinement property can be used to adaptively increase the density of the path. However, each refinement of the tool path based on the Hilbert's curve increases the tool path density in the refined region by a factor of 2 resulting in lower machining efficiency due to the increased total path length. Besides, the Hilbert's curve has an undesirable property that it leads to a path where the tool is constantly changing directions which slows down the machining process and produces large kinematics errors.

To overcome these drawbacks, adaptive SFC is used for tool path generation. It is characterized by the following useful features. First of all the adaptive SFC always follows the local optimal direction. Second, as opposed to the conventional SFC, the adaptive SFC turns only when necessary, in other words, only when the optimal direction changes. Third, the adaptive SFC eliminates the large kinematics errors and the overcuts appearing due to the sharp angular turns. Finally, local refinement of the adaptive SFC is accomplished in exactly the same fashion that the conventional SFC is refined.

A problem of adaptive SFC generation is formulated in terms of a Hamiltonian path on a grid-like graph consisting of the CC points in the (u, v) plane. The algorithm comprises three steps: (1) the grid construction, (2) the space-filling curve generation, and (3) the tool path correction.

4.3.1 Grid Construction

Consider a grid m by n obtained by overlaying two isoparametric tool paths as shown in Fig. 4.3(a-c). The isoparametric paths are generated by calculating the smallest tool path interval and using it as a constant offset to the next tool path (see Sect. 3.5). The tool path interval depends on the local surface shape, cutter shape and size, and the allowable scallop height. Besides, the machining strip widths of the two adjacent cutter paths (see Fig. 3.9) have to overlap to ensure that the machined surface error (scallop height) is within the tolerance. Techniques for calculating the machining strip width are given in Sect. 3.2.

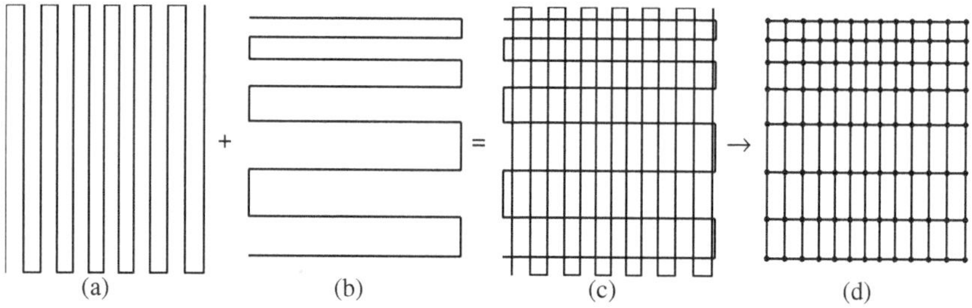

Fig. 4.3. (**a**) isoparametric tool path in the v-direction, (**b**) isoparametric tool path in the u-direction, (**c**) overlaying of two isoparametric tool paths, (**d**) the resulting grid

Furthermore, the grid is regarded as an undirected graph G where each two adjacent vertices are connected by an edge as shown in Fig. 4.3(d). The vertices of the graph correspond to initial set of CC points on the required surface whereas the distance between two connected vertices is the distance between the corresponding CC points in $\mathbb{R}^3$. Note that a cut along the path between any two connected vertices satisfies the scallop height constraint. This feature allows for the tool path optimization by means of the SFC. The SFC tool path generation algorithm is presented next.

4.3.2 Space-Filling Curve Generation

The tool path generation on the grid-like graph constructed in Sect. 4.3.1 is formulated as the Hamiltonian path problem [22]. Finding a path with the minimal length is nothing else than the traveling salesman problem. Since the problem is NP-hard [9], the algorithms for finding the optimal solution are slow and inefficient.

In this book, a simple and computationally efficient Hamiltonian path algorithm is presented. The algorithm is based on the cover and merge algorithm developed by Dafner et al. [7] for 2-dimensional image scanning. The algorithm is extended to an arbitrary rectangular grid. In addition, a correction scheme designed for the SFC tool path generation is presented (see Sect. 4.3.3).

First, all vertices are covered by small disjoint circuits. The circuits are then merged into a single Hamiltonian circuit. The initial circuits are created by constructing small rectangular cyclic paths over every 4 adjacent vertices, i.e., by connecting the vertices on even rows and columns with the vertices on odd rows and columns, respectively, as shown in Fig. 4.4(a). Besides, if m or n is odd, the virtual circuits are constructed to cover the vertices along

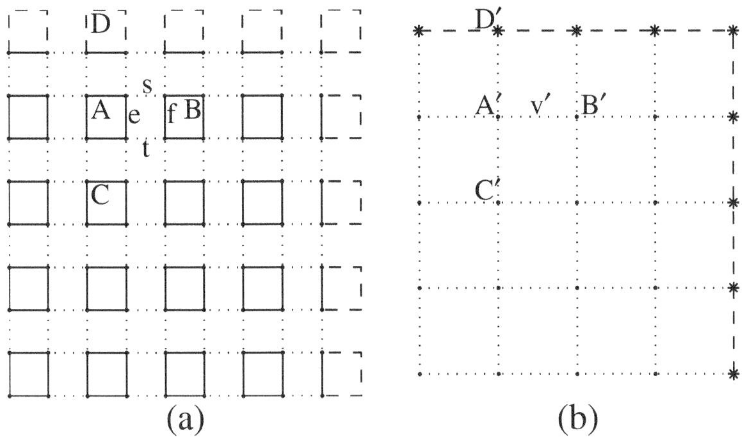

Fig. 4.4. (**a**) undirected graph G covered by initial small circuits, (**b**) dual graph G'

the boundaries as shown by the dashed lines in Fig. 4.4(a). Any two adjacent circuits can be merged into one bigger circuit. The cost of merging is defined by

$$\mathrm{Cost}(A, B) = |s| + |t| - |e| - |f|, \tag{4.3}$$

where $|e|$ denotes the distance in $\mathbb{R}^3$ between two vertices connected by an edge e.

The cost of merging two virtual circuits is set to $-\infty$, i.e., all the virtual circuits are initially merged. This is to ensure that there is no discontinuity of the tool path after removing the virtual edges from the Hamiltonian path. Furthermore, a non-virtual circuit A can be merged with a virtual circuit D only if A is merged with a non-virtual circuit C located on the opposite side. To enforce this merging dependency, the cost of merging A and D is set equal to that of merging A and C. The merging dependency is used to eliminate a possibility of an inappropriate narrow zigzag tool path with a large number of turns along the boundaries (see Fig. 4.11).

To merge all small circuits, a dual graph G' is first constructed. Each small circuit in G defines a vertex in G' and two edges s and t connecting two small circuits A and B in G define an edge v' in G' as shown in Fig. 4.4(b). Then a minimum spanning tree is constructed by iteratively merging circuits according to the cost defined by (4.3). Figure 4.5 illustrates the merging algorithm along with the construction of the corresponding minimum spanning tree. After all the circuits are merged into a Hamiltonian circuit, the tool path is generated by removing all the virtual edges, if any (see Fig. 4.5(c)). Let T be a set of edges of the minimum spanning tree. The merging algorithm is then given below.

1. Set $T = \emptyset$. Order the edges of G' in the increasing order with regard to the merging cost. In case of a tie, order the edges connecting two non-virtual vertices first.
2. Consider the first non-visited edge. Include the edge in T if and only if it does not form a circuit (in the dual graph G') with other edges from T and if and only if it does not violate the merging dependency.
3. If the edge is added to T, merge the two circuits in G corresponding to the two vertices in G' connected by the recently added edge and go to step 4, otherwise goto step 2.
4. If T includes $n-1$ edges (where n is the number of vertices of G'), stop and output the resulting Hamiltonian circuit, otherwise go to step 2.

4.3.3 Tool Path Correction

The tool path generated by the previous step requires two further modifications. First, the tool path trajectory should be modified to eliminate undercut areas. Second, the tool orientation needs to be carefully set when the tool is changing the direction.

The tool path adjustment is needed since the SFC tool path contains turns that cause the tool to miss some areas of the surface when the tool is changing the direction. At each turn, the machining strips of the two adjacent tool paths traveling in different directions might not overlap leaving an undercut or the overlap is insufficient and produces a large scallop height (see Fig. 4.6).

To eliminate this machined surface error, the tool path at the turn is modified so that the machining strip overlaps with the machining strip of the adjacent tool path. Figure 4.7 displays the two types of tool path alterations at the corner (left and right), and at the U turns. At these turns, the tool path is extended so that it overlaps with the machining strip of the adjacent tool path. The minimal extension length is

$$L_{\text{ext}} = L - w, \tag{4.4}$$

where L and w denote the tool path interval and the side machining strip width (left or right), respectively, as shown in Fig. 4.6.

The second modification is applied to the tool orientation. The tool orientation is usually set by rotating the tool about $\mathbf{y}_l$ axis toward the $\mathbf{x}_l$ axis (see the definition in Sect. 3.2) by an inclination angle λ. At sharp turns, the tool orientation changes abruptly creating large kinematics error. This kinematics error could not be reduced by merely inserting more points as usually done for tool path segmentation [2, 17]. Additionally, the tool orientation of the newly inserted CC point needs to be adjusted by interpolating the tool orientations at the two adjacent CC points. Care must be taken of when adjusting the tool orientation since changing the tool orientation could reduce the machining strip width leaving some areas undercut. Consider the sharp turn O-P-Q shown in Fig. 4.7. To make a correct turn, the $\mathbf{x}_l$ axis at the turning point

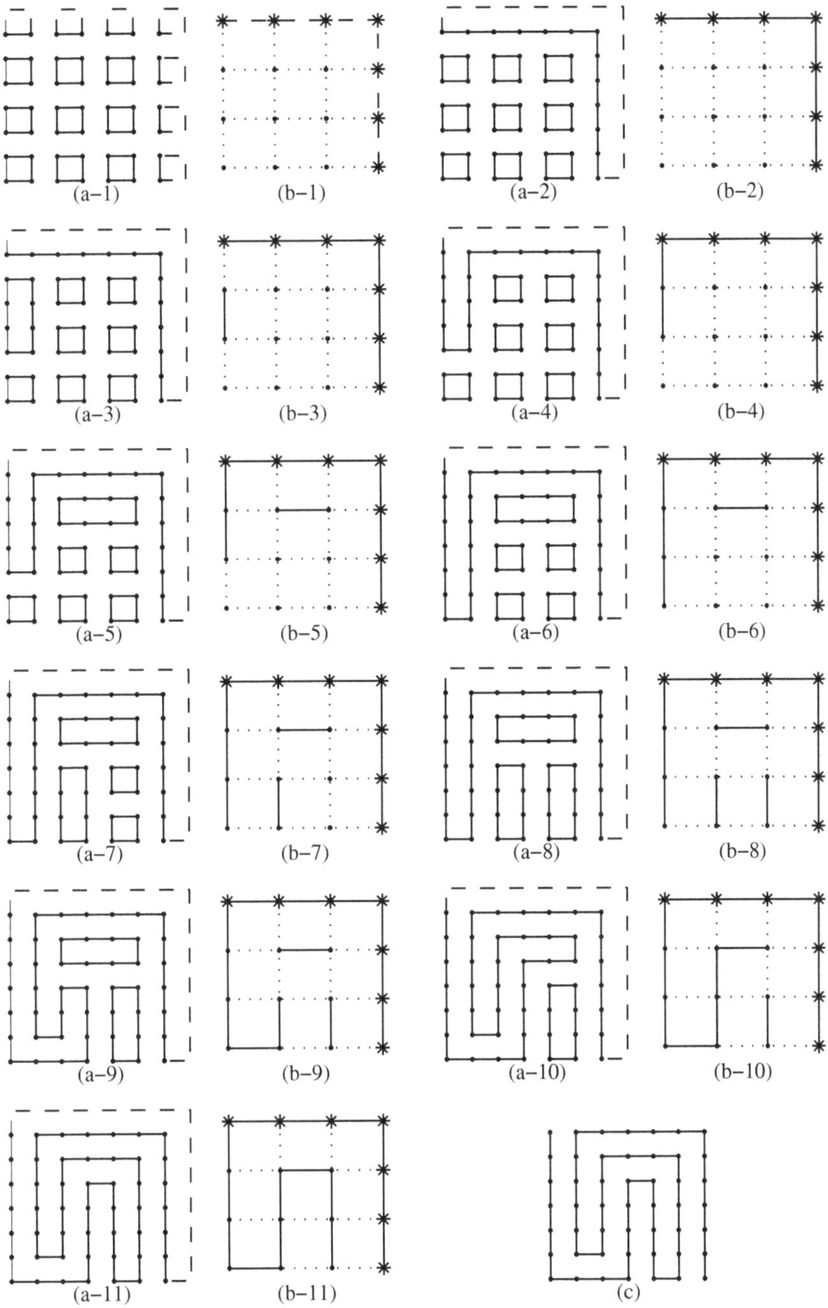

Fig. 4.5. (**a**) Hamiltonian circuit construction by merging of small circuits, (**b**) the corresponding minimum spanning tree construction, (**c**) the generated tool path

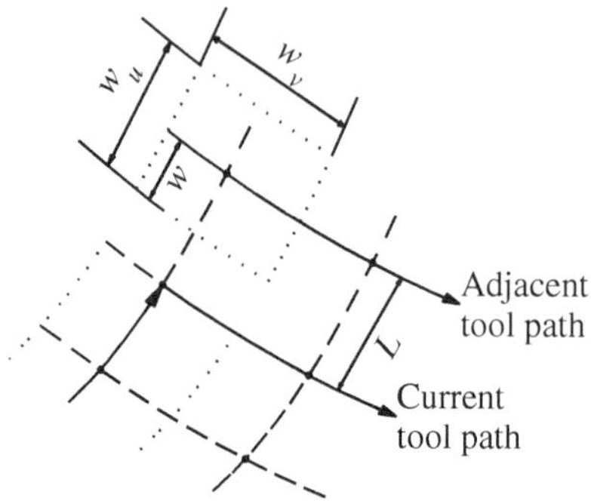

Fig. 4.6. Machining strips (*dashed lines*) on adjacent tool paths generated by using space-filling curve

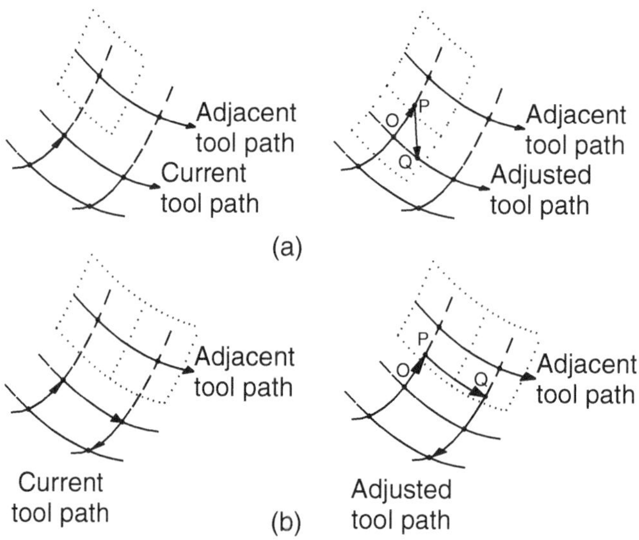

Fig. 4.7. Tool path trajectory alteration at corner turn (left or right) (**a**) and U turn (**b**)

P is redefined to be a unit surface tangent vector lying in the tool cutting direction at the previous CC point O. To reduce the kinematics error when going from point P to point Q, a new point P′ is inserted and the $\mathbf{x}_l$ axis at P′ is interpolated using the two adjacent CC points, P and Q, i.e.,

$$\mathbf{x}^*_{l,\mathrm{P}'} = \mathbf{x}_{l,\mathrm{P}'} + \mathbf{x}_{l,\mathrm{Q}'} \tag{4.5}$$

The $\mathbf{x}_l$ axis at point P′ is then found by computing the unit vector of the projection of $\mathbf{x}^*_{l,\mathrm{P}'}$ on the surface tangent plane. This process is repeated until the kinematics error between any two adjacent CC points is within the tolerance.

Figure 4.8 shows the trajectories of the effective cutting edge of the tool projected onto the **x-y** plane before and after applying the tool path correction.

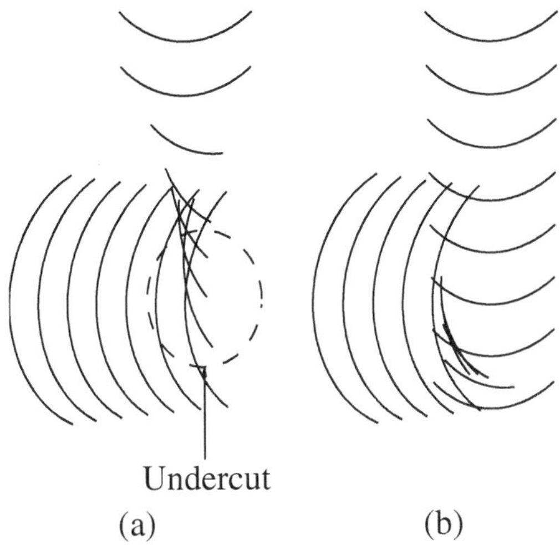

Fig. 4.8. Trajectories of the cutter's effective cutting edge (projected onto the x-y plane) before (**a**) and after (**b**) the tool path correction

Remark 4.1. The SFC tool path generation techniques are applicable to the generalized tool geometry (the so-called APT tool [3, 4]). In this case calculation of the effective cutting profile requires a numerical procedure. Once a relationship between the tool inclination and the effective cutting profile is established, the minimum tool inclination providing gouging free tool path can be computed for the generalized tool. Next, given the tool inclination, the isoparametric tool paths and hence the grid are computed. The proposed SFC generation algorithm is then executed without modifications.

4.4 Examples and Discussion

This section demonstrates the efficiency of the proposed tool path generation algorithm by examples and practical machining. In all examples, a flat-end cutter with radius $r = 3$ mm is used and the machined surface tolerance h is set to 0.01 mm.

Example 4.1. Consider a cubic Bezier surface (Fig. 4.9) consisting of both convex and concave regions. The surface is characterized by

$$P_x = \begin{bmatrix} -50 & -50 & -50 & -50 \\ -15 & -15 & -15 & -15 \\ 15 & 15 & 15 & 15 \\ 50 & 50 & 50 & 50 \end{bmatrix},$$

$$P_y = \begin{bmatrix} -50 & -15 & 15 & 50 \\ -50 & -15 & 15 & 50 \\ -50 & -15 & 15 & 50 \\ -50 & -15 & 15 & 50 \end{bmatrix}, \tag{4.6}$$

$$P_z = \begin{bmatrix} -20 & -10 & -10 & 0 \\ 0 & -10 & -10 & -20 \\ 0 & -10 & -10 & -20 \\ -20 & -10 & -10 & 0 \end{bmatrix},$$

where P_x, P_y and P_z are the x, y and z coordinates of the 16 control points. The grid constructed by overlaying two isoparametric tool paths is depicted in Fig. 4.9. The resulting tool paths are depicted in Figs. 4.10 and 4.11. The practical machining, before and after applying tool path correction, is shown in Figs. 4.12 and 4.13, respectively. Clearly, the proposed method makes it possible to eliminate the areas where the material has not been removed while maintaining the required quality of the surface. Performance of the SFC generation method versus the isoparametric method is compared in Table 4.1.

Table 4.1. Performance of the SFC tool paths versus the isoparametric tool paths in terms of tool path length

	Total path length (mm)		
Tool path	Example 4.1	Example 4.2	Example 4.3
Isoparametric in the v direction	3917.31	9397.97	7831.70
Isoparametric in the u direction	2648.12	9397.97	9036.17
SFC tool path	2637.54	7955.58	6780.84

Example 4.2. The example demonstrates the SFC being adapted to high convexity of the surface at a stationary point. Consider a symmetric bell shaped parametric surface given by

$$\begin{aligned} x &= -50 + 105u - 15u^2 + 10u^3, \\ y &= -50 + 105v - 15v^2 + 10v^3, \\ z &= -100 + 900(uv - u^2v - uv^2 + u^2v^2). \end{aligned} \tag{4.7}$$

The grid constructed by overlaying two isoparametric tool paths is depicted in Fig. 4.14. The resulting tool path is shown in Fig. 4.15. Apparently, cutting across the top of the bell requires the longest distance and yields the narrowest

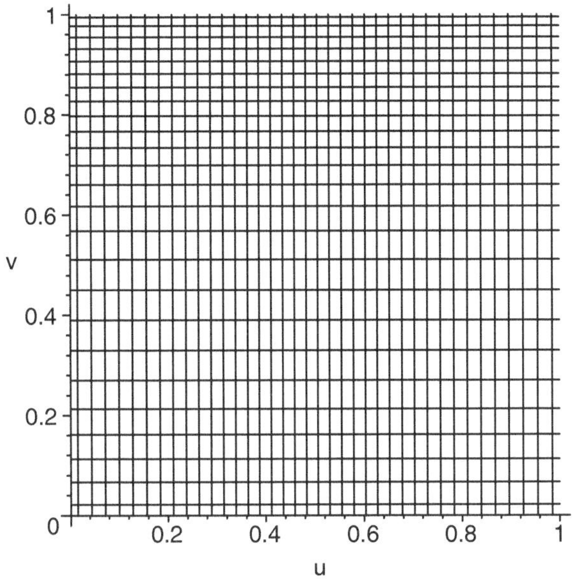

(a)

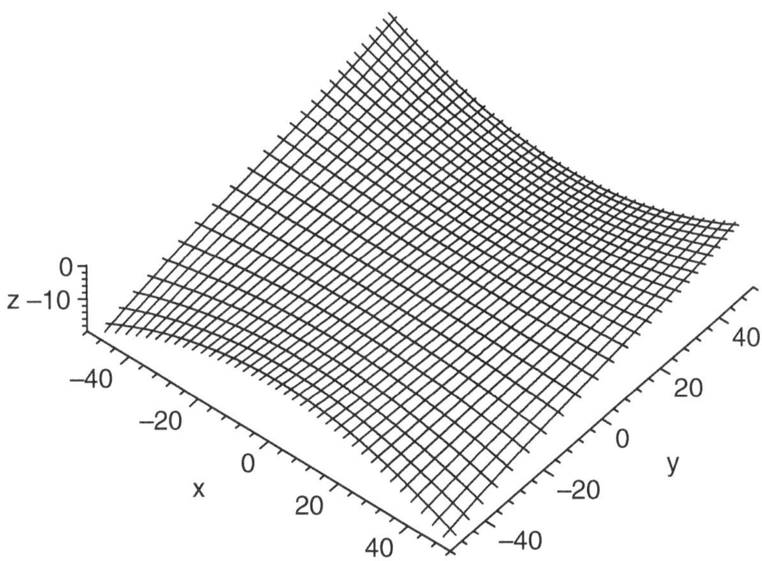

(b)

Fig. 4.9. Overlaying of two isoparametric tool paths for the surface in Example 4.1 in (u, v) domain (**a**) and in workpiece coordinate system (**b**)

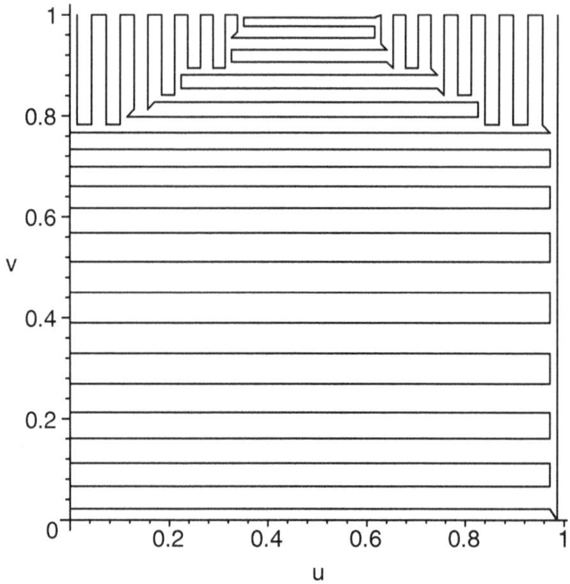

(a)

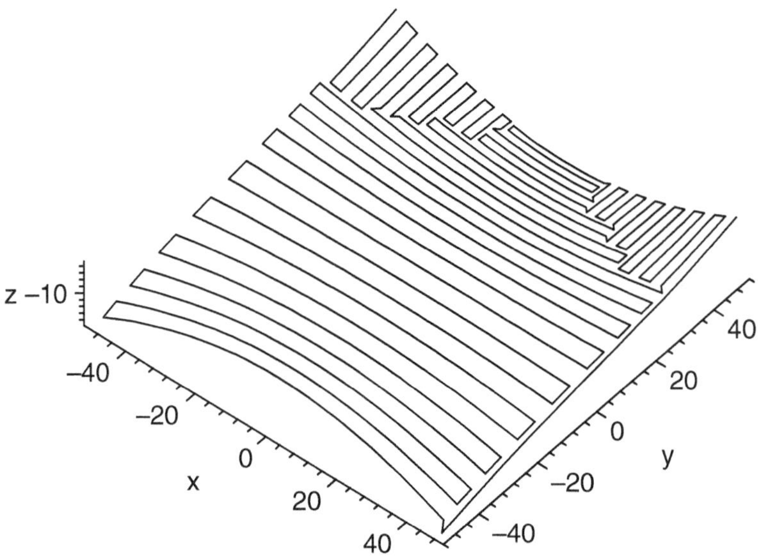

(b)

Fig. 4.10. SFC tool path for the surface in Example 4.1 in (u, v) domain (**a**) and in workpiece coordinate system (**b**)

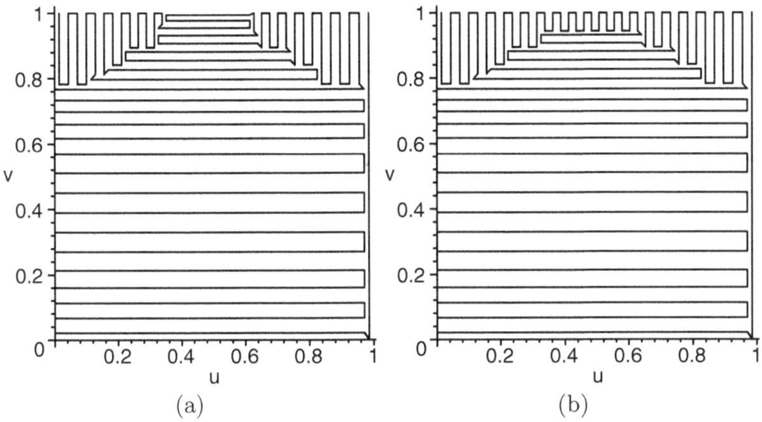

Fig. 4.11. SFC tool paths in Example 4.1 generated with (**a**) and without (**b**) merging dependency

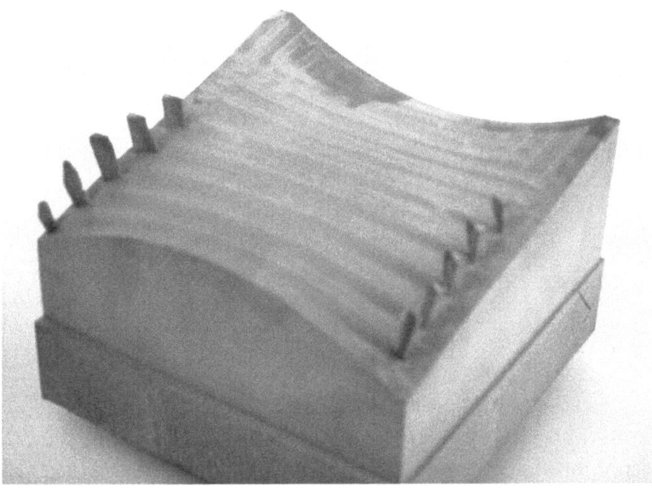

Fig. 4.12. Practical machining using the SFC tool path without correction

machining strip width due to the high convexity of the surface at the top of the bell [15]. The generated SFC tool path shown in Fig. 4.15 is adapted to the optimal direction and avoids traversing across the top of the bell.

Example 4.3. The last example demonstrates the case when the proposed technique performs extremely well compared to the traditional isoparametric tool path method. Consider a double bell surface described by the following equations:

Fig. 4.13. Practical machining using the SFC tool path with correction

$$\begin{aligned} x &= 100u - 50, \\ y &= 100v - 50, \\ z &= 400v(1 - v)(3.55u - 14.8u^2 + 21.15u^3 - 9.9u^4) - 28. \end{aligned} \tag{4.8}$$

The grid constructed by overlaying two isoparametric tool paths is depicted in Fig. 4.16. The resulting tool path is shown in Fig. 4.17. The tool path verification is performed using Unigraphics 18 and the result of the simulation is shown in Fig. 4.18.

The comparison of the tool path generation method with the traditional isoparametric method in terms of the total path length is given in Table 4.1. As seen from the results, a substantial improvement of the SFC tool path as compared with the isoparametric tool paths (Examples 4.2 and 4.3) is achieved when the required surfaces contain comparable areas characterized by different optimal cutting directions (see Figs. 4.14 and 4.16). However, the SFC tool path constructed by the proposed cover and merge algorithm does not perform well when most of the optimal cutting is in a single direction (Example 4.1, see Fig. 4.9).

Remark 4.2. The SFC tool path optimization presented above focuses on minimization of the total tool path length. However, the tool path can also be optimized with regard to some other parameters such as the time required to machine the required surface, etc. Given the spatial and angular increments, Δx_m, Δy_m, Δz_m, Δa, Δb, between two CL points and the feedrate F, the machining time t_m is calculated as follows

$$t_m = \max\{t_0, t_x, t_y, t_z, t_a, t_b\}, \tag{4.9}$$

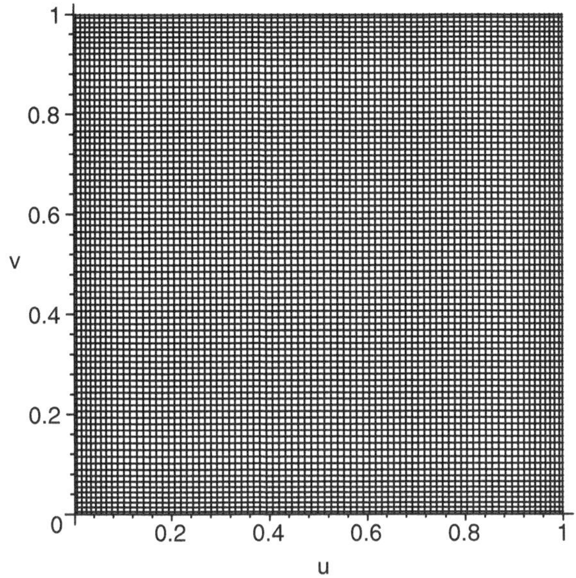

(a)

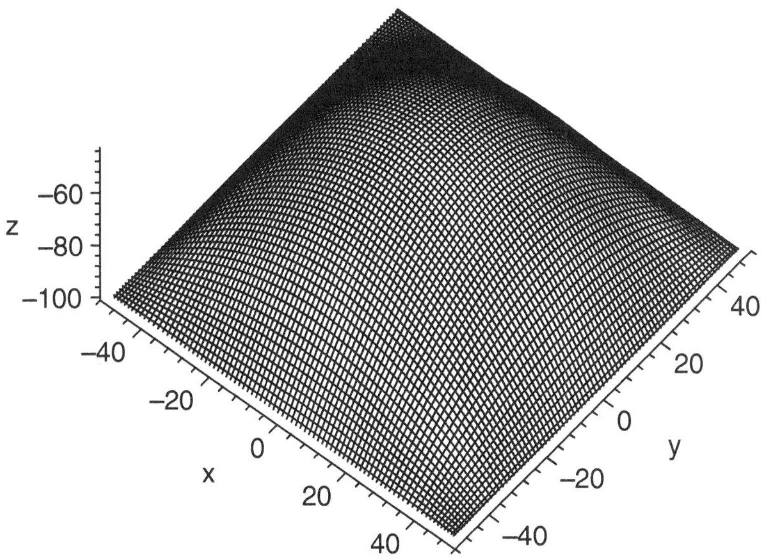

(b)

Fig. 4.14. Overlaying of two isoparametric tool paths for the surface in Example 4.2 in (u, v) domain (**a**) and in workpiece coordinate system (**b**)

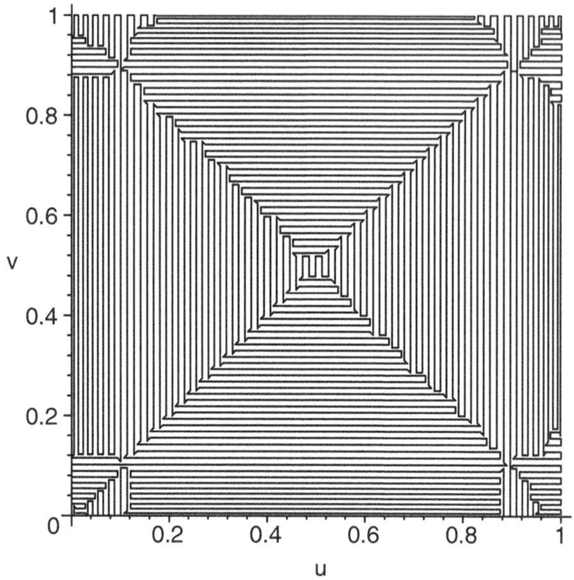

(a)

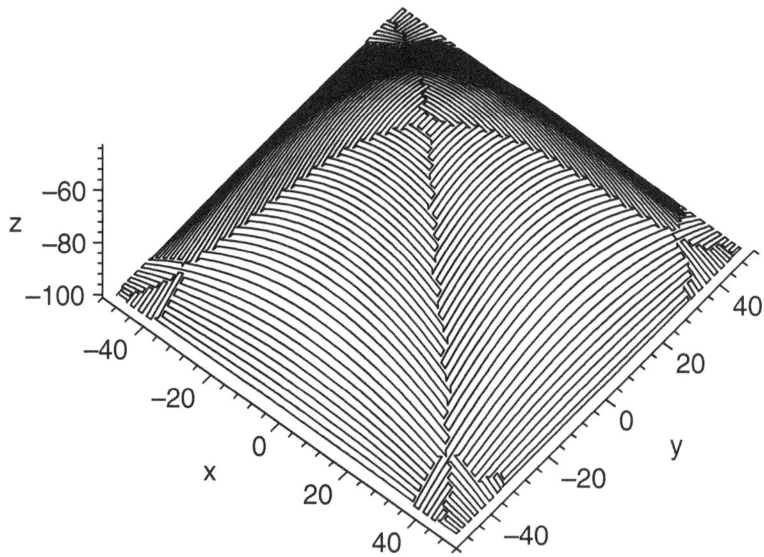

(b)

Fig. 4.15. SFC tool path for the surface in Example 4.2 in (u, v) (**a**) domain and in workpiece coordinate system (**b**)

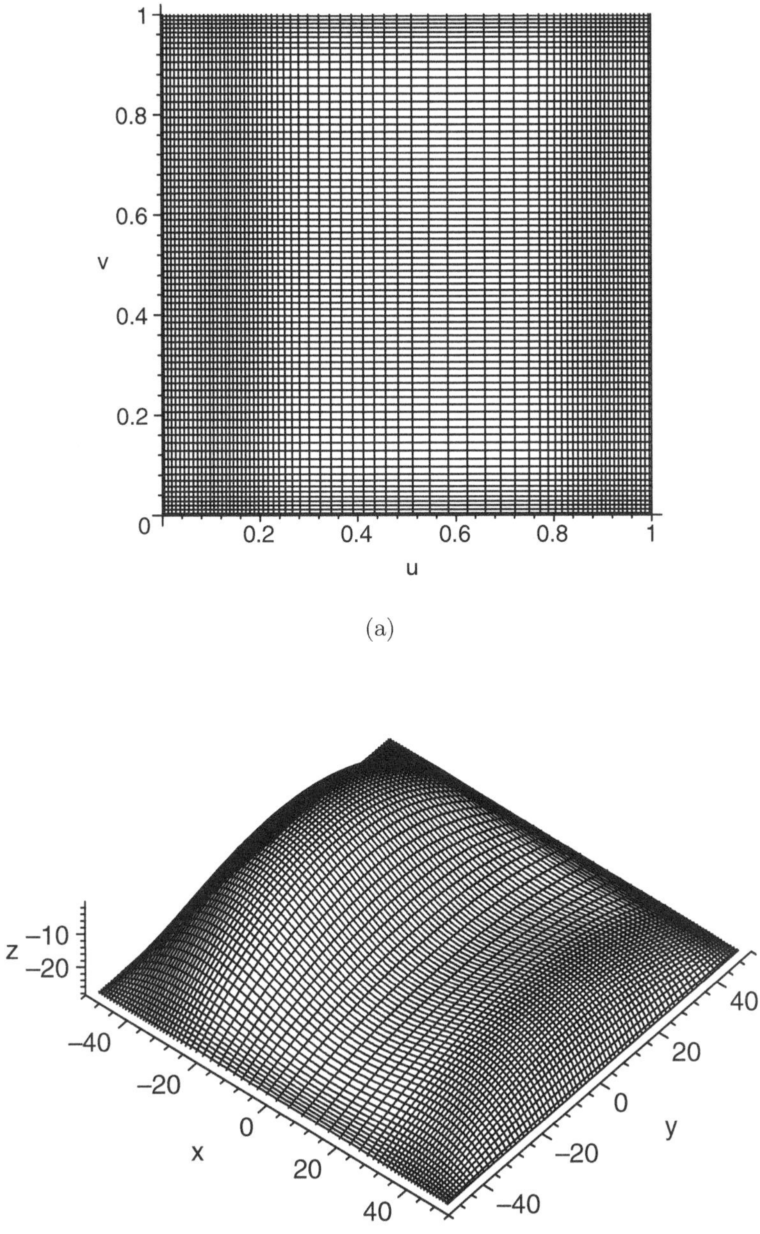

(a)

(b)

Fig. 4.16. Overlaying of two isoparametric tool paths for the surface in Example 4.3 in (u, v) domain (**a**) and in workpiece coordinate system (**b**)

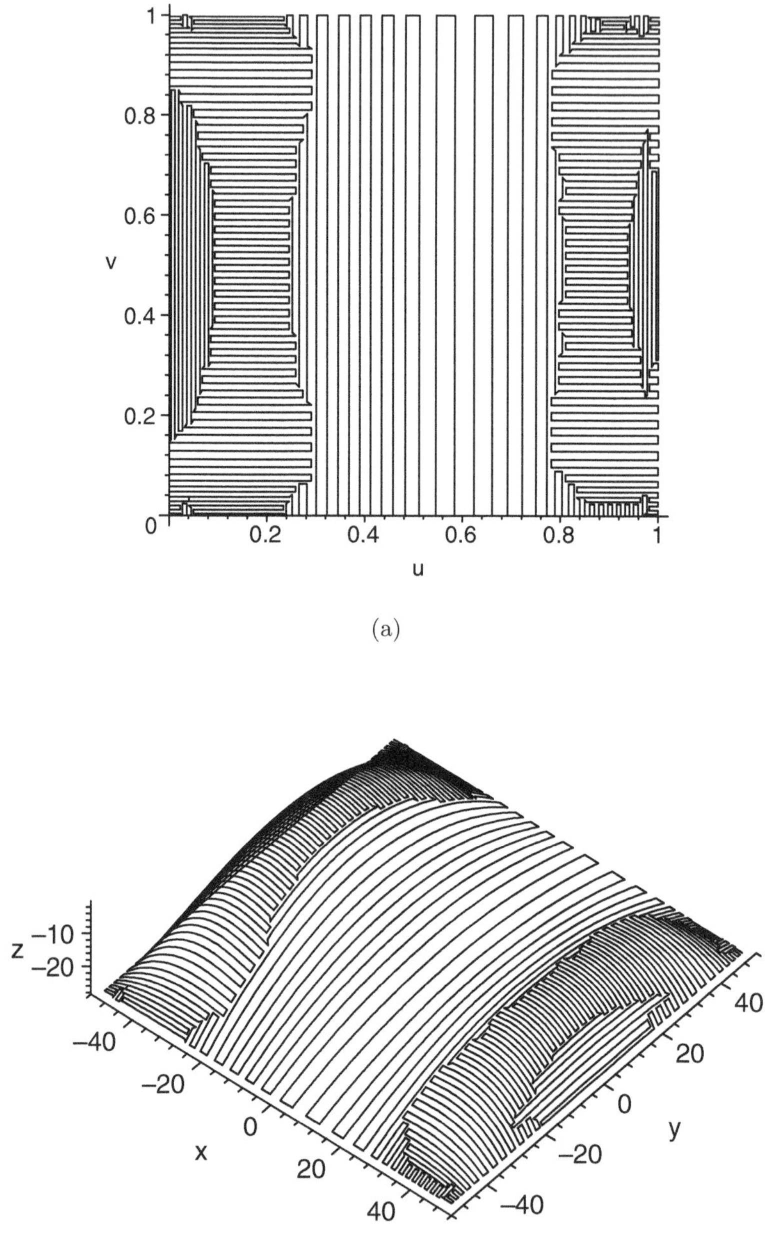

Fig. 4.17. SFC tool path for the surface in Example 4.3 in (u, v) domain (**a**) and in workpiece coordinate system (**b**)

Fig. 4.18. Simulation result of five-axis machining with SFC tool path in Unigraphics 18

where $t_0 = \frac{\Delta L}{F}$, $\Delta L = \sqrt{(\Delta x_m)^2 + (\Delta y_m)^2 + (\Delta z_m)^2}$, $t_x = \frac{\Delta x_m}{v_{x,\max}}$, $t_y = \frac{\Delta y_m}{v_{y,\max}}$, $t_z = \frac{\Delta z_m}{v_{z,\max}}$, $t_a = \frac{\Delta a}{v_{a,\max}}$, $t_b = \frac{\Delta b}{v_{b,\max}}$, and $v_{\max}$ denotes maximum velocitiy in the corresponding axis. In case of MAHO 600E, the maximum spatial and angular velocities are given by $v_{x,\max} = v_{y,\max} = v_{z,\max} = 4000$ mm/min., $v_{a,\max} = 235°/\text{sec.}$, $v_{b,\max} = 162°/\text{sec.}$

The estimated machining time for each of the tool path in Table 4.1 is given in Table 4.2. Note that a shorter tool path with many turns may take longer time than a longer tool path with fewer turns. For example, machining a simple surface such as the surface in Example 4.1 does not require many turns with sharp angle variations. Consequently, the corresponding SFC tool path outperforms the conventional isoparametric tool path in terms of the tool path length as well as in terms of the machining time. However, for a complex shaped surface such as the surface in Example 4.3, sharp turns may slow down the machining process when the angular velocity constraints are violated. For such surfaces the SFC strategy works better on the milling machines with high-speed rotational axes.

Table 4.2. Performance of the SFC tool paths versus the isoparametric tool paths in terms of estimated machining time

Tool path	Estimated machining time (seconds)		
	Example 4.1	Example 4.2	Example 4.3
Isoparametric in the v direction	4541	12342	10622
Isoparametric in the u direction	4013	12342	16744
SFC tool path	3790	11750	16414

References

[1] Asano, T., Ranjan, D., Roos, T., Welzl, E., and Widmayer, P. 1997. Space-filling curves and their use in the design of geometric data structures. *Theoretical Computer Science*, 181(1):3–15.

[2] Bohez, E. L. J. 2002. Five-axis milling machine tool kinematic chain design and analysis. *International Journal of Machine Tools and Manufacture*, 42(4):505–520.

[3] Bohez, E. L. J., Makhanov, S. S., and Sonthipermpoon, K. 2000. Adaptive nonlinear tool path optimization for 5-axis machining. *International Journal of Production Research*, 38(17):4329–4343.

[4] Chiou, C.-J. and Lee, Y.-S. 2002. A machining potential field approach to tool path generation for multi-axis sculptured surface machining. *Computer-Aided Design*, 34(5):357–371.

[5] Cole, A. J. 1987. Compaction techniques for raster scan graphics using space-filling curves. *The Computer Journal*, 30(1):87–92.

[6] Cox, J. J., Takezaki, Y., Ferguson, H. R. P., Kohkonen, K. E., and Mulkay, E. L. 1994. Space-filling curves in tool-path applications. *Computer-Aided Design*, 26(3):215–224.

[7] Dafner, R., Cohen-Or, D., and Matias, Y. 2000. Context-based space filling curves. *Computer Graphics Forum*, 19(3):209–218.

[8] Griffiths, J. G. 1994. Toolpath based on Hilbert's curve. *Computer-Aided Design*, 26(11):839–844.

[9] Hopcroft, J. E. and Ullman, J. D. 1979. *Introduction To Automata Theory, Languages, And Computation*. Addison-Wesley Longman Publishing Co., Inc., Boston, MA, USA.

[10] Kim, K. and Jeong, J. 1996. Finding tool approach directions for sculptured surface manufacture. *IIE Transactions on Design and Manufacturing*, 28(10):829–836.

[11] Koren, Y. and Lin, R.-S. 1995. Five-axis surface interpolators. *Annals of CIRP*, 44(1):379–382.

[12] Lam, W. M., Shapiro, J. M., and Sarnoff, D. 1994. A class of fast algorithms for the Peano-Hilbert space-filling curve. In *ICIP (1)*, pages 638–641.

[13] Lauwers, B., Dejonghe, P., and Kruth, J. P. 2003. Optimal and collision free tool posture in five-axis machining through the tight integration of

tool path generation and machine simulation. *Computer-Aided Design*, 35(5):421–432.

[14] Lawder, J. K. and King, P. J. H. 2001. Querying multi-dimensional data indexed using the hilbert space-filling curve. *ACM SIGMOD Record*, 30(1):19–24.

[15] Lee, Y.-S. and Ji, H. 1997. Surface interrogation and machining strip evaluation for 5-axis CNC die and mold machining. *International Journal of Production Research*, 35(1):225–252.

[16] Lin, R.-S. and Koren, Y. 1994. Real time interpolator for machining ruled surfaces. In *ASME Annual Meeting Proceedings*, pages 951–960.

[17] Makhanov, S. S., Batanov, D., Bohez, E., Sonthipaumpoon, K., Anotaipaiboon, W., and Tabucanon, M. 2002. On the tool-path optimization of a milling robot. *Computers & Industrial Engineering*, 43(3):455–472.

[18] Makhanov, S. S. and Ivanenko, S. A. 2003. Grid generation as applied to optimize cutting operations of the five-axis milling machine. *Applied Numerical Mathematics*, 46(3-4):331–351.

[19] Munlin, M.-A., Makhanov, S. S., and Bohez, E. L. J. 2004. Optimization of rotations of a five-axis milling machine near stationary points. *Computer-Aided Design*, 36(12):1117–1128.

[20] Pascucci, V. and Frank, R. J. 2001. Global static indexing for real-time exploration of very large regular grids. In *Supercomputing '01: Proceedings of the 2001 ACM/IEEE conference on Supercomputing (CDROM)*, pages 2–2.

[21] Rao, A. and Sarma, R. 2000. On local gouging in five-axis sculptured surface machining using flat-end tools. *Computer-Aided Design*, 32(7):409–420.

[22] Roberts, F. S. 1984. *Applied combinatorics*. Prentice-Hall, Inc., Upper Saddle River, NJ, USA.

[23] Sagan, H. 1994. *Space-Filling Curves*. Springer-Verlag, New York.

[24] Schamberger, S. and Wierum, J.-M. 2005. Partitioning finite element meshes using space-filling curves. *Future Generation Computer Systems*, 21(5):759–766.

[25] Song, Z. and Roussopoulos, N. 2000. Using Hilbert curve in image storing and retrieving. In *MULTIMEDIA '00: Proceedings of the 2000 ACM workshops on Multimedia*, pages 167–170.

[26] Velho, L. and Gomes, J. 1991. Digital halftoning with space filling curves. In *SIGGRAPH '91: Proceedings of the 18th annual conference on Computer graphics and interactive techniques*, pages 81–90.

[27] Yang, K.-M., Wu, L., and Mills, M. 1988. Fractal based image coding scheme using Peano scanning. In *Proceedings of 1988 IEEE International Symposium on Circuits and Systems*, pages 2301–2304.

[28] Zhang, Y. and Webber, R. E. 1993. Space diffusion: an improved parallel halftoning technique using space-filling curves. In *SIGGRAPH '93: Proceedings of the 20th annual conference on Computer graphics and interactive techniques*, pages 305–312.

5

Tool Paths in Adaptive Curvilinear Coordinates

5.1 Introduction

Space-filling curves tool path generation technique presented in Chap. 4 has been designed for surfaces represented in such a way that the parametric coordinates are changing within a rectangular region. Once a minimal machining strip is evaluated, constructing a basic grid for the SFC generation in the rectangular region is trivial (see Sect. 4.3.1). However, such a grid is often inefficient since a small step between the tracks could be required only in certain areas. The rectangular grid is also inefficient in the case of a complex boundary of the so-called trimmed surfaces. These surfaces are characterized by the boundaries created by intersections with other surfaces. The complex boundaries also occur in the case of pocket milling when the parametric region includes internal boundaries around one or several pockets. From the viewpoint of computational mathematics the above are classic situations when a numerically generated adaptive curvilinear grid should be introduced. The grid may be easily converted to the zigzag tool path or replace the basic grid required at the first step of the SFC tool path generation method.

In this chapter we introduce and analyze the both possibilities. The first approach is a modification of a classic grid generation based on the Euler-Lagrange equations for a functional which represents desired properties of the grid such as smoothness, adaptivity to the boundaries and to a certain weight (control) function [15, 81]. Therefore, Sect. 5.3 presents a concatenation of the grid generation approach and the tool path optimization. In this framework, proposed in [49, 51], the weight function is represented by the kinematics error. The minimization is subjected to constrains relevant to the heights of scallops between the successive tool tracks.

However, the above techniques have a number of drawbacks. In particular, they may converge slowly for complicated constraints. Besides, the approach requires equal number of the cutter contact points on each track of the tool. Therefore, if the kinematics error changes sharply from track to track, the method may require an excessive number of points.

Section 5.4 introduces another modification of the grid refinement which fits better to the framework of tool path optimization. The method does not require equal number of points on each track. It automatically evaluates the number of the required grid lines. As opposed to the first approach, where the weight function represents either the kinematics error or an estimate of the kinematics error (such as the surface curvature or the rotation angles), the proposed algorithm iteratively constructs an adaptive control function designed to represent the scallop height constraints. This important modification makes it possible to consider an arbitrary number of points along the tool tracks. Consequently, the kinematics error is reduced by means of inserting additional points along the resulting curvilinear coordinates[1]. In other words, this approach replaces the scallop constraints by a weight function and then treats the kinematics error independently. Additionally, instead of the Winslow functional the new optimization is based on the harmonic functional derived from the theory of harmonic maps [25, 36]. The functional not only provides the smoothness and the adaptivity but under certain conditions guarantees the numerical convergence as well. Finally, this approach merges with the SFC techniques. In this case, the grid is not converted to the tool path directly. Instead, it becomes the basic grid required for SFC generation which replaces the rectangular grid (see Chap. 4). With this modification, the SFC tool path can be constructed to machine surfaces with complex irregular boundaries, cuts off, pockets, islands, etc. Besides, the adaptive grid allows to efficiently treat complex spatial variability of the constraints in such a way that the SFC is being created on a grid having the small cells only where necessary.

5.2 A Historical Note on Grid Generation

Grid generation has been developed as a sub-discipline of computational fluid dynamics (CFD), heat and mass transfer and structural analysis. Nowadays, grid generation is employed in many other areas including CAD/CAM, mechanical engineering and manufacturing.

Structured grid generation has its roots in the US in the works of Winslow with Lawerence Livermore National Lab in the late 1960s [81]. About the

[1] Curvilinear coordinates are a coordinate system based on some transformation of the standard coordinate system. For example, consider the 2D case, instead of Cartesian coordinates x and y one can use e.g. p and q where $p = p(x, y)$ and $q = q(x, y)$. The level curves of p and q in the (x, y) plane, as well as those of x and y in the (p, q) plane are in general curved. It is required that the transformation is locally invertible at each point. This means that a point can be converted in one coordinate system to its curvilinear coordinates and back. Depending on the application, a curvilinear coordinate system may be simpler than the Cartesian coordinate system. This also has consequences that the concepts in vector calculus can be expressed in Cartesian coordinates and also in curvilinear coordinates [1].

same time the structured grid generation was proposed in Russia by Godunov and Prokopov with Keldysh Institute of Applied Mathematics in Moscow and Computing Center of the Siberian Branch of the USSR Academy of Sciences [32]. Some ideas, foreshadowed the modern grid generation, were presented by Tikhonov and Gorbunov [75], Sidorov [67] and Bahvalov [6].

The unifying idea of grid generation is based on an estimate of a certain weight (cost) function combined with the equi-distribution principle which imposes such a distribution of the nodes that a product of the grid spacing and the weight function remains constant through the entire region. Some pioneering papers introducing this idea are [17, 22, 35, 58, 62, 72].

A variety of weight functions suitable for solutions of problems occurring in CFD has been proposed in [2, 3, 5, 13–15, 21, 23, 24, 27, 28, 52, 53, 55, 74, 79, 80].

A general viewpoint on the problem is represented by the concept of the so-called *informative-computational space* [2, 40, 54, 72]. The grid is characterized by the so-called *elliptic property* interpreted as *intercommunication of observers.* A motion of an observer to a suitable position implies that his neighbors follow him. The term *elliptic* originates from the fact that such intercommunication can be simulated by elliptic partial derivative equations such as the Laplace equation. A number of semi-heuristic methods employing various interpretations of the informative computational space have been proposed by Baker [7], Berger and Oliger [8], Blacker and R.J. [10], Blacker and Stephenson [11], Dannenhoffer [20], Frey and Field [30], Jiang and Carey [43], Kennon and Anderson [44], Kennon and Dulikravich [45], Pardhanani and Carey [56], Rank and Babushka [60], Samareh-Abolhassani and Stewart [65], Soni [68], Vabistchevitch [78].

Furthermore, the term *structured* means that the curvilinear grid is topologically equivalent to a rectangular grid. In other words, the curvilinear grid can be obtained by applying certain kind of deformations to the rectangular one. Eventually this idea evolved into the idea of constructing a grid which represents a discrete version of a one-to-one mapping of the so-called *physical region*, where the curvilinear grid is defined, onto a parametric square (or cube in 3D) defined in the so-called *computational region*, where the image of the grid becomes rectangular [32, 33, 73].

It turned out that construction of such grids can be performed by means of variational methods which employ Euler equations to optimize a weighted average of the grid quality measures. The grid quality measures include a number of criteria such as smoothness of the grid, orthogonality, adaptation to a certain weight function in accordance with the equi distribution principle, the total length of the grid lines, cell aspect ratio and many others [64].

Considerable effort has been spent on elimination of self-intersected and twisted cells and nodes located outside the prescribed region when the mapping becomes unacceptable [37]. Such grid is called the *degenerate* grid. An efficient approach to construct non-degenerate grids is based on the concept of *orthogonal grid* proposed and analyzed in [4, 14, 15, 59, 66, 70]. An orthogonal grid generator involves a measure of orthogonality represented by a dot product between the vectors tangential to the grid lines [13, 15, 63].

Another approach to construct a grid without degenerate cells is based on *multi-block grid strategies* employing a decomposition of the complex-shaped regions into blocks [32, 73, 76]. However, the discontinuities at the boundary between blocks is the chief drawback of the method [18, 69].

The concept of *convex grid* has been proposed by Ivanenko [38]. The grid which consists of convex quadrilaterals ensures a non-degenerate grid at the continuous as well as at the discreet level [36, 61]. Furthermore, the concept of non-degenerate grids can be represented in the framework of the harmonic maps [26]. The use of the harmonic maps as a suitable grid generation strategy was proposed by Dvinsky [25] and Ivanenko [36]. The map constructed on the surface of the graph of the weight function provides the required adaptation of the grid in accordance with the equi distribution principle. The advantage that the theory features over other grid generation strategies is that in 2D it guarantees (under some conditions) the existence and the uniqueness of the map. Besides, harmonic grid generation has been shown to be parameterization-independent [46], namely, that the resulting grid in physical space is independent of the choice of parameterization for the physical domain.

As far as the tool path generation is concerned, construction of the tool paths of industrial milling machines in the framework of grid generation technologies was first introduced in [49] and developed in [50]. The grid generation techniques are surprisingly well-adapted to tool path optimizations. As a matter of fact, the concept of a grid refinement contains almost all the main ingredients for tool path planning, such as: grid adaptation to the regions of large milling errors, possibility to easily construct curvilinear versions of the conventional zigzag and spiral patterns and adaptation to constraints related to the tool diameter and the scallop height. Moreover, in contrast to the standard techniques characterized by a local error estimate, grid generation deals with a global spatial error and consequently adapts all the CL points simultaneously.

The ideas were further developed in [51], specifically for five-axis machining and in the framework of the harmonic maps. Bieterman and Sandstrom [9] suggested a similar approach based on the Laplacian grid generator independently.

It should be noted that application of grid generation for tool path optimization has not been fully exploited. For example, unstructured grids are generally far more flexible and robust when applied to complicated regions [12, 31, 34, 57]. However, application of such grids for tool path generation is an open problem.

Note that our references are related to selected pioneering papers introducing basic concepts of structured grid generation and harmonic maps. As far as the entire scope of modern grid generation methods is concerned, the interested reader should refer to excellent texts of Carey [16], Farrashkhalvat and Miles [29], Liseikin [46, 47], Steinberg [71], Thompson et al. [73], Tucker [77].

Finally, in spite of success of grid generation in many areas, the process is still not made easy enough and automatic. The generation of grids could become the most time consuming part of calculations requiring hours and even days of calculations.

5.3 Variational Grid Generation for Tool Path Optimization

5.3.1 Preliminary Examples

A preliminary example is a surface having sharp variations along a sinus shaped curve Fig. 5.1. The corresponding curvilinear grid is depicted in Fig. 5.2. The tool will move along one family of the coordinate curves. It is plain that such curves are more appropriate for cutting and require the same data structures as those employed for rectangular grids.

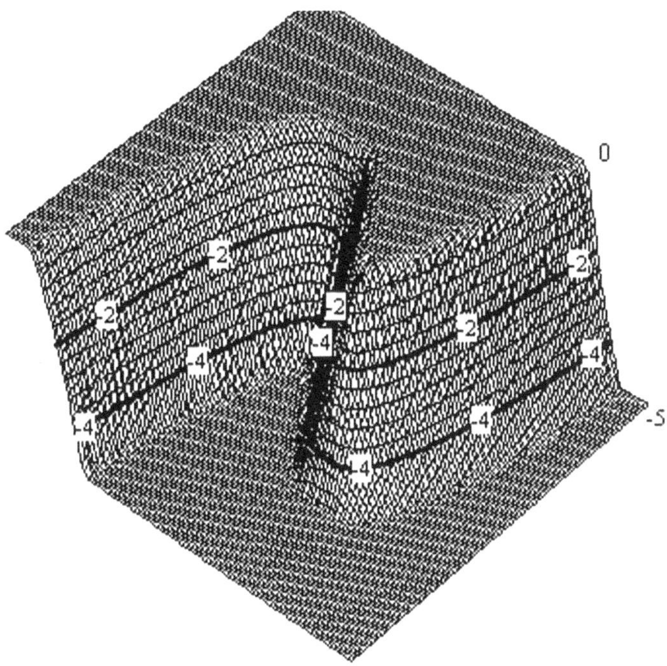

Fig. 5.1. Surface with a curvilinear zone of large gradients

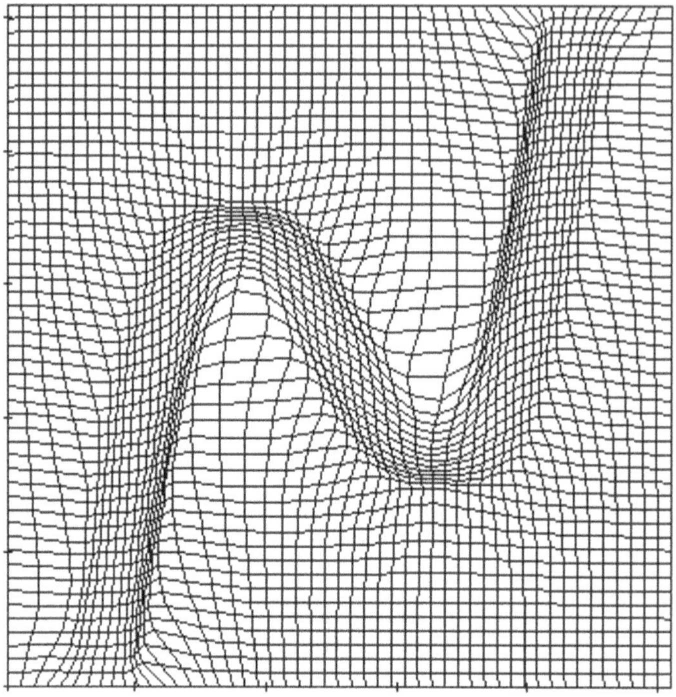

Fig. 5.2. Curvilinear grid which can be converted into a tool path to machine the surface in Fig. 5.1

Furthermore, the grid generation techniques are applicable to generate a tool path in the case of complex boundaries such as the case of a complex shaped domain in Fig. 5.3. Jeong and Kim [41] address this domain as an example of complex pocket milling which may not be solved by means of a regular zigzag pattern. However, the generated grid shows that the techniques enable us to simultaneously generate appropriate curvilinear zigzags and curvilinear spirals. The grid is well adapted to the internal and external boundaries. Besides, the flexibility of the grid generation approach allows for further adaptation to the regions requiring more accurate machining.

5.3.2 Variational Method and Functionals

In this section we present basic principles of variational grid generation methods and show how these techniques can be applied to tool path optimization. As far as the grid generation techniques are concerned, we follow the classical variational approach of Winslow [81] and Brackbill and Saltzman [15]. The tool path optimization based of adaptive grids follows [49, 51].

Recall that $S \equiv S(u,v) \equiv (x(u,v), y(u,v), z(u,v))$ denotes a surface to be machined. As usual u and v are the parametric variables. Consider a set

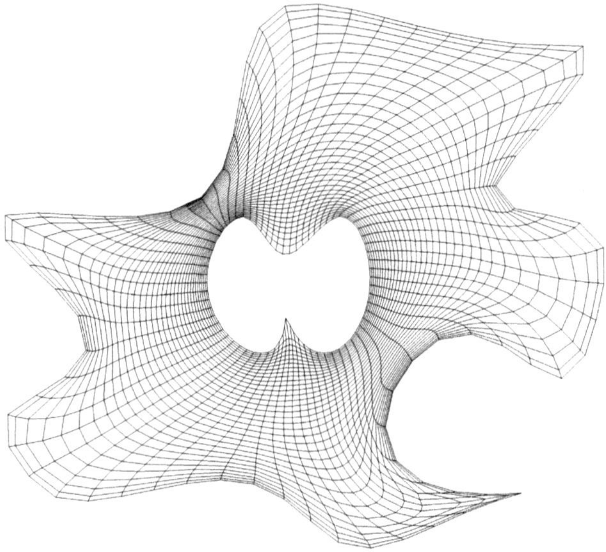

Fig. 5.3. A grid which can be converted into a tool path for a complex shaped region

of cutter location points $\{u_{ij}, v_{ij}\}, 0 \leq i \leq N_\xi, 0 \leq j \leq N_\eta$ arranged as a curvilinear grid. Mathematically, it means that $\{u_{ij}, v_{ij}\}$ is a discrete analogy of a mapping from the *computational region* $\Delta = \{0 \leq \xi \leq N_\xi, 0 \leq \eta \leq N_\eta\}$ onto a *parametric region* defined in the parametric coordinates u, v. In other words there exists a pair of functions $u(\xi, \eta)$, $v(\xi, \eta)$ such that a rectangular grid $\{i, j\}$ being fed to $u(\xi, \eta)$, $v(\xi, \eta)$ becomes $\{u_{ij}, v_{ij}\}$ (Fig. 5.4).

Recall that the general tool path optimization problem is given by

$$\min_{S, \Pi, \mathfrak{T}, \mathfrak{M}} C,$$

where C denotes a criteria vector. The vector may include the kinematics error, the length of the path, the negative of the machining strip, the machining time, etc. Furthermore, Π is the tool path, $\mathfrak{M}$ is a set of the parameters related to the setup and configuration of the machine, and $\mathfrak{T}$ is the parameters of the tool. Usually, the optimization is subjected to constraints which may include the scallop height, gouging avoidance constrains, etc.

Note that optimization criteria and constraints may be interchangeable depending on the machining problem. For example, the minimization can be performed with regard to the kinematics error subject to the scallop height constraint. Alternatively, the scallops can be minimized subject to a prescribed maximum error. A weighted sum of the scallops and the tool tip errors can be also considered as a single criteria.

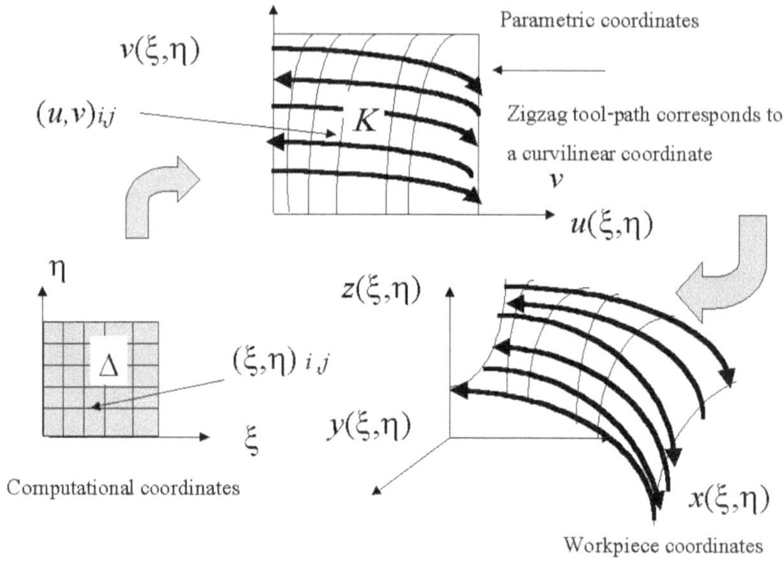

Fig. 5.4. Tool path as a mapping from the computational to the parametric (physical) domain

In this section we consider the criteria vector C, consisting of a single element given by $C = ||\epsilon||$, where $\epsilon = |S(\xi,\eta) - T(\xi,\eta)|$ and $T(\xi,\eta)$ is a machined surface. An approximation of T can be obtained using a variety of techniques including solid modeling, the sweep volume methods, etc (see Introduction).

As an example of a simple approximation consider $T(\xi,\eta)$ composed from subsurfaces (patches) $T_{i+1/2,j+1/2}(\xi,\eta)$ spanned onto a grid-cell $\{(u,v)_{i,j}, (u,v)_{i+1,j}, (u,v)_{i,j+1}, (u,v)_{i+1,j+1}\}$, where $T_{i+1/2,j+1/2}(\xi,\eta)$ is obtained by linear interpolation between the tool trajectories (Fig. 5.5).

Since the tool path Π is now represented by the grid of points $\{u_{ij}, v_{ij}\}$, we arrive at the following optimization.

$$\min_{u_{ij},v_{ij}} ||\epsilon||. \tag{5.1}$$

Furthermore, keeping (5.1) in mind, let us introduce grid generation techniques based on variational principles. Following [15, 81], the required grid is a discretized solution of a minimization problem given by

$$\min\,(F_s + \lambda_v F_v + \lambda_o F_o), \tag{5.2}$$

where

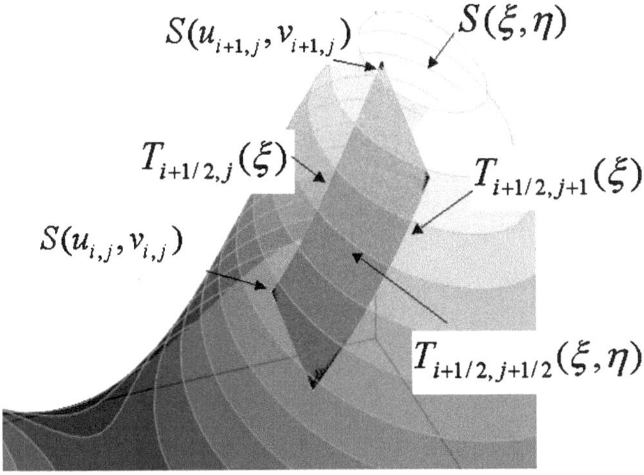

Fig. 5.5. Constructing surface $T(\xi, \eta)$ from patches

$$F_s \equiv \iint_{\Delta} [(\Delta\xi)^2 + (\Delta\eta)^2]\, dudv, \tag{5.3}$$

$$F_v \equiv \iint_{\Delta} J\epsilon\, dudv, \tag{5.4}$$

$$F_o \equiv \iint_{\Delta} \Delta\xi \cdot \Delta\eta\, dudv. \tag{5.5}$$

$\Delta \equiv (\frac{\partial}{\partial u}, \frac{\partial}{\partial v})$, J denotes the Jacobian of the mapping given by $J = \frac{\partial u}{\partial \xi}\frac{\partial v}{\partial \eta} - \frac{\partial v}{\partial \xi}\frac{\partial v}{\partial \eta}$, λ_v, λ_o are the calibration parameters. Furthermore, F_s is the Winslow functional which measures the smoothness of the mapping. A mapping that minimizes the functional produces a grid which does not have discontinuities or sharp corners and fits to the boundary of the physical region.

F_o is the so-called orthogonality functional. Its minimum $F_o = 0$ is reached when all the conjugate lines of the grid are orthogonal. The functional can be used as an additional measure which improves the quality of the grid i.e. eliminates twisted and degenerate cells. However, in the context of tool path generation F_o does not represent an important machining criteria.

Finally, F_v is the measure of the equi distribution with regard to weight function ϵ. The product of the area of the grid cell (which is nothing else than the discretized Jacobian J) and the weight function ϵ should be constant through the entire grid.

This is where the problem of the error minimization meets the grid generation. Since the error tends to zero as the area of the grid cell tends to zero, the grid generator adapts the cells in such a way that the error is reduced or even minimized while keeping the basic grid structure. However, F_v can not be minimized individually because the solution is not unique or may not exist. For example, $J = \infty$ minimizes F_v, however, this mapping is singular and, therefore, does not produce a curvilinear grid. Let us omit the orthogonality functional and consider $F = F_s + \lambda_v F_v$. (Further analysis of the orthogonality functional can be found in [15, 42]).

Changing variables in F_s and F_v yields

$$F_s = \iint\limits_{\Delta} J^{-1}\left[\left(\frac{\partial u}{\partial \xi}\right)^2 + \left(\frac{\partial u}{\partial \eta}\right)^2 + \left(\frac{\partial v}{\partial \xi}\right)^2 + \left(\frac{\partial v}{\partial \eta}\right)^2\right] \mathrm{d}\xi \mathrm{d}\eta, \tag{5.6}$$

$$F_v = \iint\limits_{\Delta} J^2 \epsilon \, \mathrm{d}\xi \mathrm{d}\eta. \tag{5.7}$$

There is a variety of ways to proceed from that point. One of them is employing the Euler equations to minimize the weighted sum of the functionals $F_s + \lambda F_v$. The Euler equations are given by

$$\frac{\partial F}{\partial x} - \frac{\partial}{\partial \xi}\frac{\partial F}{\partial x_\xi} - \frac{\partial}{\partial \eta}\frac{\partial F}{\partial x_\eta} = 0,$$

$$\frac{\partial F}{\partial y} - \frac{\partial}{\partial \xi}\frac{\partial F}{\partial y_\xi} - \frac{\partial}{\partial \eta}\frac{\partial F}{\partial y_\eta} = 0.$$

Next, the partial derivatives should be replaced by the finite differences and the system of the finite difference equations should be solved numerically.

Alternatively, functionals F_s and F_v are approximated by quadratures I_s and I_v . The resulting discrete functional $I = I_s + \lambda_v I_v$ should be minimized using the conditions

$$\frac{\partial I}{\partial u_{ij}} = 0, \frac{\partial I}{\partial v_{ij}} = 0.$$

The both methods produce the same system of finite difference equations when the standard second order approximations are being used for the first method ($\frac{\partial^2 u}{\partial \xi^2} \approx u_{i+1,j} - 2u_{i,j} + u_{i-1,j}$, $\frac{\partial u}{\partial \eta} \approx \frac{u_{i+1,j} - u_{i-1,j}}{2}$, etc.) and when the standard trapezoidal rule with $\Delta\xi = \Delta\eta = 1$ is applied to discretize F_s and F_v for the second method.

It is not hard to demonstrate that the system of finite difference equations is given by

$$\begin{aligned} \tilde{R}_u &\equiv a_1 u_{\xi\xi} + a_2 u_{\xi\eta} + a_3 u_{\eta\eta} + c_1 v_{\xi\xi} + c_2 v_{\xi\eta} + c_3 v_{\eta\eta} + J^2 \epsilon_v, \\ \tilde{R}_v &\equiv b_1 u_{\xi\xi} + b_2 u_{\xi\eta} + b_3 u_{\eta\eta} + a_1 v_{\xi\xi} + a_2 v_{\xi\eta} + a_3 v_{\eta\eta} + J^2 \epsilon_u, \end{aligned} \tag{5.8}$$

where the subscripts ξ, η, u and v correspond to the finite differences at (ξ_i, η_j) and (u_{ij}, v_{ij}), respectively.

The coefficients at the derivatives of the second order are given by

$$a_k = a_{s,k} + \lambda_v a_{v,k},\ b_k = b_{s,k} + \lambda_v b_{v,k},\ c_k = c_{s,k} + \lambda_v c_{v,k},\ k = 1, 2, 3,$$

$$a_{s,1} = -A\alpha,\ b_{s,1} = B\alpha,\ c_{s,1} = C\alpha,$$

$$a_{s,2} = 2A\beta,\ b_{s,2} = -2A\beta,\ c_{s,2} = -2C\beta,$$

$$a_{s,3} = -A\gamma,\ b_{s,3} = B\gamma,\ c_{s,3} = C\gamma,$$

$$A = u_\xi v_\xi + u_\eta v_\eta,\ B = v_\xi^2 + v_\eta^2,\ C = u_\xi^2 + u_\eta^2,$$

$$a_{v,1} = -u_\eta v_\eta,\ b_{v,1} = v_\xi^2,\ c_{v,1} = u_\eta^2,$$

$$a_{v,2} = u_\xi v_\eta + u_\eta v_\xi,\ b_{v,2} = -2v_\xi v_\eta,\ c_{v,2} = -2u_\xi u_\eta,$$

$$a_{v,3} = -u_\xi v_\eta,\ b_{v,3} = v_\xi^2,\ c_{v,3} = u_\xi^2,$$

Let us now evaluate the machining strip in the midpoint between two consecutive tracks of the tool η_j and η_{j+1}. Consider the flat-end cutter and for simplicity suppose that the tool is aligned with the normal to the surface. Consider a vertical plane through points $S(\xi_i, \eta_j)$ and $S(\xi_i, \eta_{j+1})$. In this plane we approximate the corresponding section of the surface by a circle with the radius R, where R is the radius of the curvature in that direction. A simple geometric reasoning (Fig. 5.6) yields

$$w = 2|R|\sqrt{\frac{2|R|h + h^2}{|R| + h}} \tag{5.9}$$

Additionally the tool radius r must satisfy $r \geq 2\sqrt{2|R|h + h^2}$. Note that, if the tool must be inclined, (5.9) must be replaced by a formula which includes the effective cutting radius of the tool r_e given by (3.36) (see Sect. 3.3). For instance, for $h \ll r_e$, $h \ll R$, Lo [48] suggested a formula similar to (3.34) given by

$$w = \sqrt{\frac{8hr_eR}{R - r_e}}.$$

In order to construct a tool path simultaneously adapted to the milling errors ϵ and to the constraints related to the prescribed scallop height, we introduce the following conditions:

$$D_{i,j+1/2} \geq 0, \tag{5.10}$$

where $D_{i,j+1/2} = w^2_{i,j+1/2} - d^2_{i,j+1/2}$ and d denotes the distance between the points on the adjusted curves η_j and η_{j+1} on the surface $S(u, v)$ given by

$$d_{i,j+1/2} = \sqrt{(x_{i,j+1} - x_{i,j})^2 + (y_{i,j+1} - y_{i,j})^2 + (z_{i,j+1} - z_{i,j})^2} \tag{5.11}$$

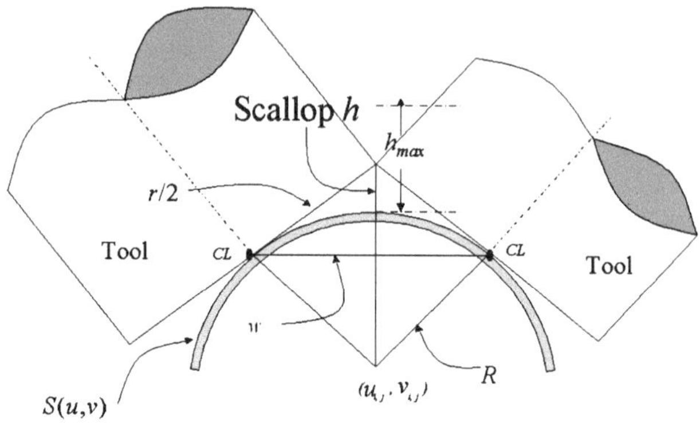

Fig. 5.6. The scallop height evaluation for convex surfaces

and w is evaluated by (5.9).

Thus, we arrived at the following problem: solve equations (5.8) subject to constrains (5.10) to find a minimizer of functional I. In order to solve the *constraint* minimization, we introduce a penalty function $p(D)$. It is a convex decreasing function such that, $p(D) = 0$ if $D \geq 0$ and $p(D) \to \infty$ when $D \to -\infty$.

Next, we define a discrete penalty functional given by

$$I_p = \lambda_p \sum_{i,j} \lambda_{ij} p(D_{i,j+1/2}),$$

where $\{\lambda_{ij}\}$ are the penalty coefficients and λ_p is the weight coefficient. $p(D), D \in (-\infty, 0)$, is a convex decreasing function, $p(D) \to \infty$ if $D \to -\infty$.

Next, the derivatives of $\lambda_p I_p$ with regard to u_{ij}, v_{ij} are added to the left hand side of the above finite difference equations, namely,

$$R_u = \tilde{R}_u + \lambda_p \frac{\partial I_p}{\partial u_{ij}} = 0, \; R_v = \tilde{R}_v + \lambda_p \frac{\partial I_p}{\partial v_{ij}} = 0.$$

Example 5.1. Suppose that $S(u,v) \equiv (x,y,z) = (u,v,u^2+v^2)$. Consider a penalty function given by $p(D) = [\min(0,D)]^2$. Then for $D < 0$ we have

$$\begin{aligned} p(D_{i,j+1/2}) &= D_{i,j+1/2}^2, \\ &= [w^2 - \Big((u_{i,j+1} - u_{i,j})^2 + (v_{i,j+1} - v_{i,j})^2 \\ &\quad + (u_{i,j+1}^2 + v_{i,j+1}^2 - u_{i,j}^2 - v_{i,j}^2)^2\Big)]^2. \end{aligned}$$

Differentiation with regard to $u_{i,j}$ yields a penalty term given by

$$\begin{aligned}\frac{\partial I_p}{\partial u_{i,j}} =& 2D_{i,j+1/2}\big[2(u_{i,j+1}-u_{i,j}) \\ &\quad + 4(u_{i,j+1}^2+v_{i,j+1}^2-u_{i,j}^2-v_{i,j}^2)u_{i,j}\big]\lambda_{i,j+1/2} \\ &- 2D_{i,j-1/2}\big[2(u_{i,j}-u_{i,j-1}) \\ &\quad + 4(u_{i,j}^2+v_{i,j}^2-u_{i,j-1}^2-v_{i,j-1}^2)u_{i,j}\big]\lambda_{i,j-1/2}.\end{aligned}$$

Note that $p(D)$ is not differentiable at $D=0$. Therefore, one has to perform an appropriate regularization near $D=0$. For instance,

$$p_{\mathrm{reg}}(D) = \begin{cases} p(D), & \text{if } D \geq \epsilon, \\ H(D), & \text{if } 0 < D < \epsilon, \\ 0, & \text{otherwise,} \end{cases}$$

where $H(D)$ is a Hermite polynomial satisfying $H(\epsilon)=p(\epsilon)$, $\dfrac{dH(\epsilon)}{dD}=\dfrac{dp(\epsilon)}{dD}$, $H(0)=0$.

The corresponding system can be solved by quasi Newtonian iterations given by

$$u_{i,j}^{n+1} = u_{i,j}^n - \left[\tau\left(R_u\frac{\partial R_v}{\partial v} - R_v\frac{\partial R_u}{\partial v}\right)\left(\frac{\partial R_u}{\partial u}\frac{\partial R_v}{\partial v} - \frac{\partial R_v}{\partial u}\frac{\partial R_u}{\partial v}\right)^{-1}\right]_{i,j},$$

$$v_{i,j}^{n+1} = v_{i,j}^n - \left[\tau\left(R_v\frac{\partial R_u}{\partial u} - R_u\frac{\partial R_v}{\partial u}\right)\left(\frac{\partial R_u}{\partial u}\frac{\partial R_v}{\partial v} - \frac{\partial R_v}{\partial u}\frac{\partial R_u}{\partial v}\right)^{-1}\right]_{i,j},$$

where n is the iteration index and τ is the iteration parameter. The penalty coefficients are updated by the iterative procedure

$$\lambda_{i,j+1/2}^{l+1} = \begin{cases} \lambda_{i,j}^l + \delta\lambda_{i,j}^l, & \text{if } D_{i,j+1/2} < 0, \\ \lambda_{i,j}^l, & \text{otherwise,} \end{cases}$$

where l is the index of the penalty iterations and $\delta\lambda_{i,j}^l$ denotes the corresponding increment. In order to improve the stability of the algorithm, we use the following linear relaxation

$$\begin{aligned}(u_{i,j}^{n+1})_{\mathrm{new}} &= u_{i,j}^{n+1}(1-\theta) + u_{i,j}^n\theta, \\ (v_{i,j}^{n+1})_{\mathrm{new}} &= u_{i,j}^{n+1}(1-\theta) + u_{i,j}^n\theta,\end{aligned}$$

where $0 \leq \theta \leq 1$.

5.3.3 The Harmonic Functional

The sum of two functionals F_s and F_v in (5.2) can be replaced by a single functional given by

$$I = \int \frac{(u_\xi^2 + u_\eta^2)(1 + f_u^2) + (v_\xi^2 + v_\eta^2)(1 + f_v^2) + 2 f_u f_v (u_\xi v_\xi + u_\eta v_\eta)}{(u_\xi v_\eta - u_\eta v_\xi)\sqrt{1 + f_u^2 + f_v^2}} d\xi d\eta. \tag{5.12}$$

(5.12) is called the harmonic functional and is a generalization of Winslow functional to the case of *error surfaces* described explicitly by $f(u, v)$. The subscripts u, v, ξ, η denote partial derivatives and f is the weight function.

The harmonic functional can be derived from the theory of harmonic maps [25, 36]. It has been proven that the functional minimizes an *energy* of mapping (see details [42]) and produces a grid adapted to the regions of large gradients of f. Note that if $f_u \equiv f_v \equiv 0$, then the harmonic functional (5.12) becomes the Winslow functional (5.3). It is important though to understand that I adapts the grid to the gradients of f rather than to f itself. Therefore, it works differently as compared with F_v. Besides, minimization of (5.12) could be 10 times more computationally expensive that minimization of (5.2) [19]. However, (5.12) has some points in its favor. In particular, it is possible to construct a computational procedure which (under certain conditions) always converges to a non degenerate grid [42]. The constraint minimization of (5.12) subject to the scallop height conditions can be performed by the penalty type techniques similar to those presented in the preceding section. The required modifications and the approximation of the harmonic functional are given in Sect. 5.4.

5.3.4 Examples of the Tool Path Optimization

This section illustrates the performance of the proposed procedures by educational examples.

Example 5.2. **Adaptation to a curvilinear zone**

The example illustrates convergence of the tool path generation method for a workpiece having curvilinear boundaries and a sin-shaped zone of large milling errors depicted in Fig. 5.7.

For simplicity let us specify constraints (5.9)-(5.10) without defining the actual surface. Suppose we cut the surface by a flat-end cutter and suppose that the minimum radius of the curvature of the surface $R = 45$. We will require that the maximum scallops do not exceed $h = 0.1$ mm. Substituting $R = 45$, $h = 0.1$ into (5.9) yields $d < 1.25$. We will impose this constraint for the entire region, although at many points the minimal distance will be underestimated. Furthermore, the error is *emulated* by (see Fig. 5.7)

$$\epsilon = \exp\left(-\left|y - 0.5\,(L_y - 6)\sin\left(\frac{\pi x}{L_x - 0.5}\right) + 0.5\right|\right).$$

Figure 5.7 displays generation of the curvilinear tool path satisfying the prescribed constraints generated by $p(D) = [\min(D, 0)]^2, \delta\lambda = 1, \lambda_v = 10, \lambda_p = 0.9$.

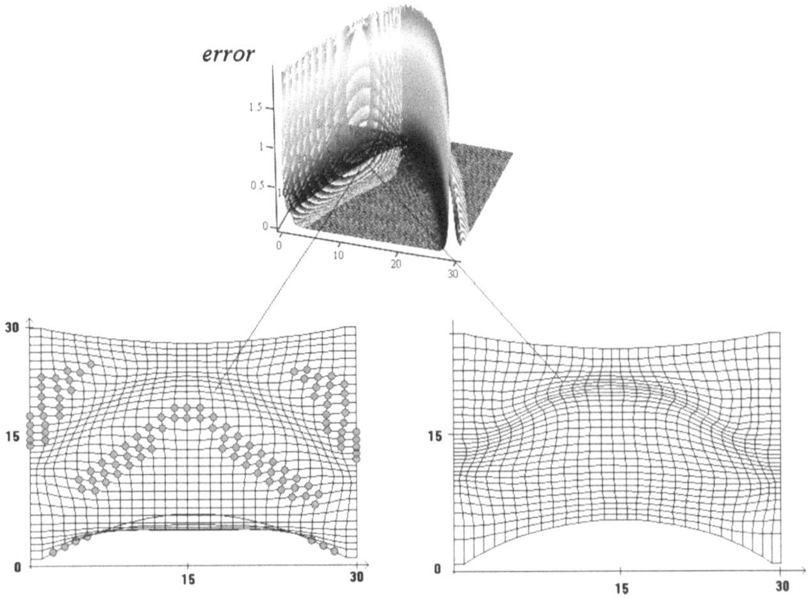

Fig. 5.7. Tool path generation with adaptation to a weight function and the boundary

Table 5.1 illustrates the convergence of the iterations and reveals that $\lambda_p = 0.9, \upsilon = 0.25$ in this particular case lead to the best convergence rate. The symbol # indicates a situation when the iterations do not have a limit and either fluctuate or reach the computer infinity. Such situation is called divergence.

Table 5.1. Convergence of the method λ_p versus θ

$\lambda_p \backslash \theta$	0.10	0.25	0.50	0.75
0.01	836	900	1001	1200
0.10	301	320	380	402
0.25	115	54	331	380
0.50	201	56	256	371
0.90	90	<u>52</u>	210	330
1.00	82	58	201	301
1.25	78	63	180	270
1.50	#	#	#	265
1.75	#	#	#	#

Example 5.3. **A spiral tool path embedded into a zigzag tool path**

This example demonstrates how to produce a tool path composed from patches corresponding to different types of motion. The generated tool path in Fig. 5.8 has been adapted to the curvilinear boundaries and to three zones of large milling errors located inside the circular region and at the left part of the domain. The surface curvature and the scallop height have been set up as in Example 5.2, *artificially*, that is, without employing an actual surface.

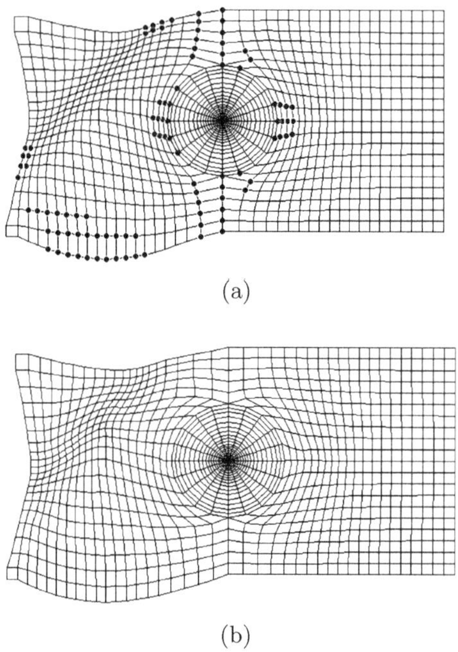

Fig. 5.8. Grids obtained by constraint minimization (**a**) and unconstrained minimization (**b**)

The grid in Fig. 5.8(a) is constructed by an unconstrained minimization. Clearly, the grid is unacceptable since it has many *bad points*, where the constraints are not satisfied. The maximum scallop height is about 50 times more than the prescribed value. Furthermore, in order to prevent a too small distance between grid points located in the irrelevant regions it is often useful to impose a lower bound by adding a second constraint given by

$$d - w' \geq 0, \text{if } \epsilon < \epsilon_1,$$

where ϵ_1 is a small positive constant. This helps to avoid degenerated grids.

The composed tool path generated with $\epsilon_1 = 10^{-5}, w = 1.6, w' = 0.6$ shown in Fig. 5.8(b) is adapted subject to the both prescribed constraints with $p(D) = [\min(D, 0)]^2, \delta\lambda = 1, \lambda_v = 10, \lambda_p = 0.25$.

Example 5.4. **A Bezier surface**

The constraint minimization techniques could produce a grid which actually does not decrease ϵ (even increase it). This is because our model implies that the accuracy is determined by the pair (ϵ, h) rather than solely by ϵ. Consequently, if the minimization does not decrease of the error, the user should either increase the number of the tool tracks or apply a trial-and-error compromise between ϵ and h. The following example illustrates the trial-and-error procedure. We consider a Bezier surface (Fig. 5.9) characterized by the following control points.

$$B_x = \begin{pmatrix} 0 & 0 & 0 & 0 \\ 12 & 12 & 12 & 12 \\ 24 & 24 & 24 & 24 \\ 36 & 36 & 36 & 36 \end{pmatrix}, B_y = B_x^T, B_z = \begin{pmatrix} -16 & -8 & -6 & 0 \\ -8 & -6 & 0 & -2 \\ -6 & 0 & -2 & 0 \\ 0 & -2 & 0 & 0 \end{pmatrix}.$$

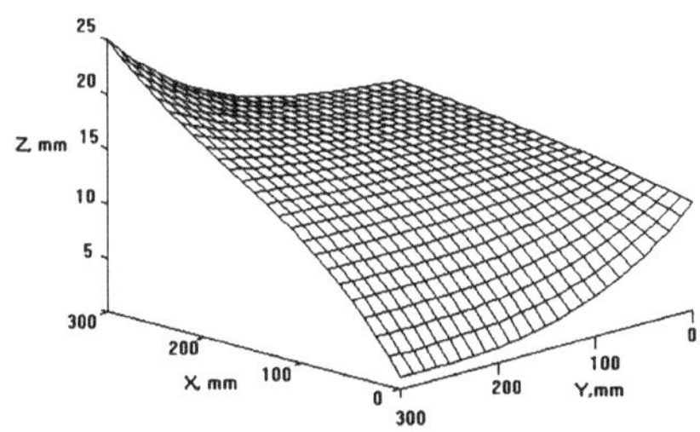

Fig. 5.9. A test Bezier surface

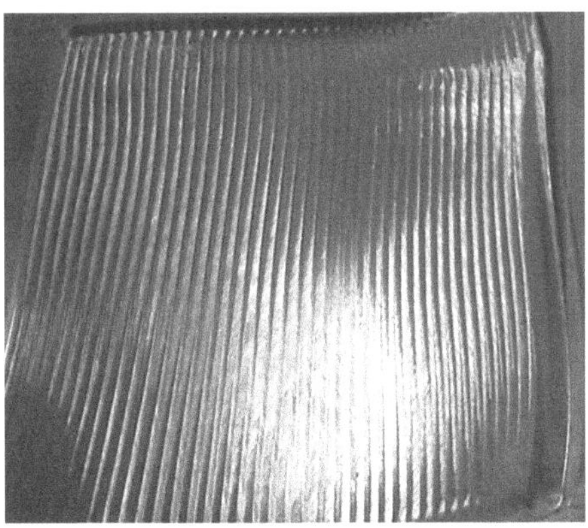

Fig. 5.10. The machined Bezier surface

Figure 5.10 displays an actual machining. A conventional tool path is shown in Fig. 5.11(a). The bad points are indicated by circles sized proportionally to the distance from the constraints, whereas points with large ϵ-values are indicated by boxes with the size proportional to ϵ. The resulting $\epsilon = 0.08$.

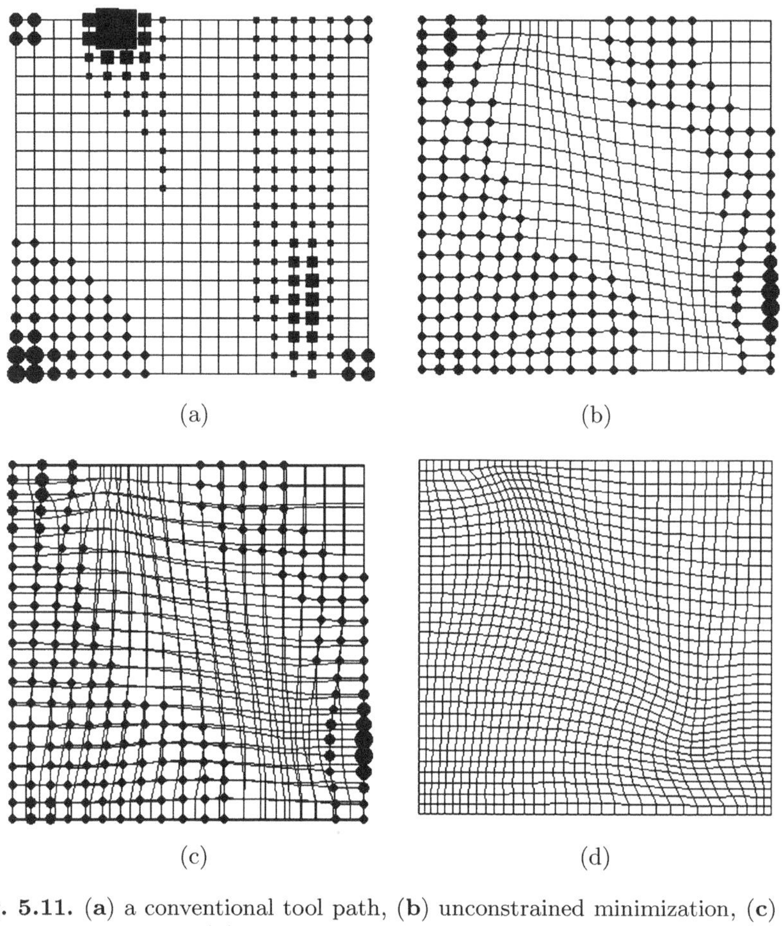

Fig. 5.11. (a) a conventional tool path, (b) unconstrained minimization, (c) constrained minimization, (d) constraint minimization on a grid 35×35

A grid generated by the unconstrained minimization (Fig. 5.11(b)) yields $\epsilon = 0.04$. Finally, the constrained minimization Fig. 5.11(c) produces $\epsilon = 0.085$. In order to construct an appropriate tool path, either the maximum allowed scallop height or the number of the grid points has to be increased.

We analyze the both proposed options. Increasing h from 0.01 to 0.04 yields an adapted tool path satisfying the prescribed scallop constraints $\epsilon = 0.041$. On the other hand, an adapted tool path having 35×35 CL points (Fig.

5.11(d)) yields $\epsilon = 0.021$, whereas a conventional 35×35 tool path still does not satisfy the scallop constraints. Observe the possibility to adapt a tool path if only a cross-feed adjustment is required. A suitable adaptation could then be solely performed by penalty functions, i.e. by a discrete functional given by $I_s + \lambda_p I_p$. Alternatively the scallop constraints may be included into F_v or even into the harmonic functional (5.12). The last possibility will be discussed in the next section. The orthogonality measure could also be included into the algorithm to ensure against degenerate grids and divergence.

A series of test surfaces has been tested by means of the proposed algorithm. Table 5.2, demonstrates an average accuracy increase ranging between 32 and 43%.

Table 5.2. Accuracy of the machined surface

Grid size	Scallop (mm)	improvement % of conventional
10×10	3.60	34 / 0.2600
20×20	1.80	41 / 0.0930
30×30	1.20	34 / 0.0580
40×40	0.90	36 / 0.0410
60×60	0.60	32 / 0.0260
80×80	0.45	40 / 0.0180

It is plain that the constraints related to scallops may substantially affect the solution. Besides, the solution to the minimization problem is not unique or may not exist. In practice, the initial grid typically does not satisfy the constraints and there is no prior knowledge whether such a grid exists. Consequently, the convergence is analyzed by means of numerical experiments. Note, that the adaptive approach allows nor a closed form estimate of the number of required tracks neither an estimate of the number of the CL points belonging to one track. Therefore, we developed a realistic rule applicable to the majority of practical situations. If a number of the CL points not satisfying the scallop constrains is more than 2/3 of the total number of points than the adaptation is not possible or requires a very large number of iterations. In other words, the computational efficiency of the algorithm depends on the number of points admissible for re-distribution. If this number is small than, first of all, the error can not be substantially reduced, secondly, the optimal grid may not exist at all. Even if such a grid exists, the iterations may fail to converge. Table 5.3 illustrates computational complexity of the algorithm for a 40×40 zigzag tool path constructed for the Bezier surface. The accuracy increase characterized by the ratio $\epsilon_R = 100\epsilon/\epsilon_C$, where ϵ_C is the error on the rectangular pattern, τ_C the computational time on a PC (Pentium 4), I' the number of the required iterations, N_B the number of the *bad points*, where the constraint $h = 0.01$ is violated, from the initial rectangular grid, # indicates divergence. Observe that $h = 0.015$ mm leads to a substantial accuracy increase of about 40 % with regard to the rectangular tool path, whereas

$h = 0.013$ mm yields only a 22% increase. $h = 0.001$ mm and $h = 0.012$ mm actually increase the error. It should be noted that $h = 0.015$ mm seems to be the most suitable since the scallop height and the error are approximately in the same magnitude. The calculations reveal that h and ϵ varying in the same magnitude often lead to the minimum number of iterations. It means that appropriate scallop heights could be taken as fractions of ϵ corresponding to the conventional tool path.

Table 5.3. Convergence of the algorithm h versus ϵ

Scallop h (mm)	Computational time τ_C (min)	Iterations I'	Error ϵ	Error relative to conventional ϵ_R	Number of bad points N_B
0.0001	#	#	#	#	1132
0.0010	32	1710	0.094	200	500
0.0012	11	550	0.047	105	92
0.0130	9	421	0.032	78	61
0.0150	4	250	0.024	59	25
0.0170	3	200	0.020	55	0

The grid generation algorithms applied to tool path generation offer a number of benefits such as the global two-dimensional adaptation to the regions of large milling errors. As opposed to many tool path generation methods (see Introduction) the grid searches for a global minimum adapting all the CL points simultaneously. Of course such an adaptation leads to certain disadvantages such as possible divergence, large computational time, etc. However, solutions obtained by the grid generation technologies can hardly be obtained by the local methods which may require astronomical times to reach the same optimum. Furthermore, the grid generation provides a straightforward control of the constraints related to a scallop height and capabilities to generate a tool path for workpieces with complex shaped boundaries for trimmed surfaces

Finally, the grid generation technologies are particularly suitable to be integrated with the conventional CAD/CAM software in so far that the algorithms deal with a curvilinear versions of standard zigzag or spiral patterns. Therefore, such modifications require only an additional mapping obtained by means of the grid generation procedures.

5.4 Application of Harmonic Functional to Tool Path Generation

Recall that $S \equiv S(u, v) \equiv (x(u, v), y(u, v), z(u, v)$ denotes the required surface given in the workpiece coordinates, where u and v are parametric variables. Consider a set of CC points, $\{(u, v)_n\}, 1 \leq n \leq N_n$, being a discrete analogy of a mapping from the computational region on the plane (ξ, η) into the physical region defined in the parametric coordinate system (u, v) (see Fig. 5.4).

A more general case is a regular or structured grid which consists of curvilinear quadrilaterals such as the grid shown in Fig. 5.12 or a block-structured grid in Fig. 5.13. In both cases, the grid is not topologically equivalent to a rectangular Cartesian grid. These two cases invoke mapping of the unit square in the plane (ξ, η) onto each quadrilateral in the (u, v) plane individually. The block-structured grids can also be used to construct the SFC whereas the use of a general irregular grid to construct an appropriate SFC lies out of the scope of this book.

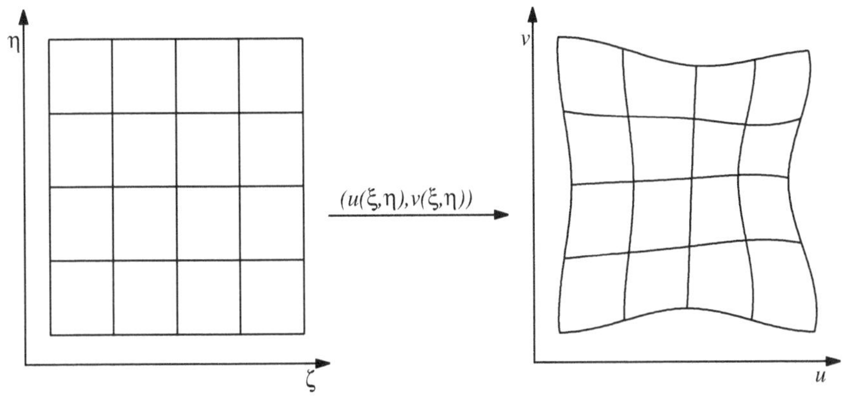

Fig. 5.12. Example of regular grid

The problem of grid generation can be treated as a discrete analog of the problem of finding functions $u(\xi, \eta)$ and $v(\xi, \eta)$. It can be shown that if a smooth mapping of one domain onto another with one-to-one mapping between boundaries possesses a positive Jacobian, then such a mapping will be one-to-one [1]. Hence, the curvilinear coordinate system constructed in (u, v) domain will be non-degenerate if the Jacobian of the mapping $u(\xi, \eta)$, $v(\xi, \eta)$ is positive, i.e.,

$$J = u_\xi v_\eta - u_\eta v_\xi > 0. \tag{5.13}$$

A number of techniques for grid generation have been developed [47]. In general, the methods of grid generation for structured grid can be classified into three basic groups, namely, the algebraic methods, the differential methods and the variational methods. In the algebraic approach the interior points of the grid are computed through various forms of interpolation or special functions. Differential methods generate grid points based on the solution of elliptic and parabolic equations. In variational methods, the grid is generated by optimization of grid quality properties such as non-degeneracy, smoothness, uniformity, near-orthogonality, or adaptivity (see, for examples, [15, 19, 37, 38]).

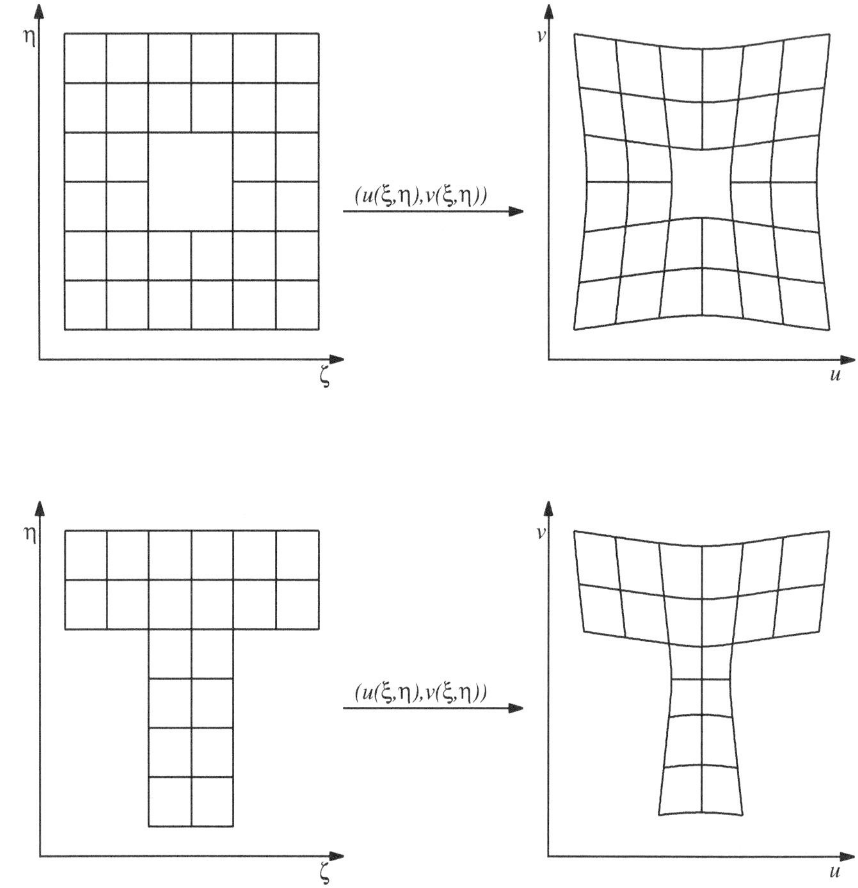

Fig. 5.13. Examples of block-structured grids

This section presents the grid generation technique based on variational grid generation method of Ivanenko [39]. The grid is considered as a discrete version of a mapping $\{u(\xi, \eta), v(\xi, \eta)\}$, where the pair $\{u, v\}$ minimizes a Dirichlet functional [39] (5.12).

The discrete version of the solution is the so called adaptive-harmonic grid characterized by small cells in the regions of large gradients of the control function $f(u, v)$ (see, for example, Fig. 5.14). The grid can be also interpreted as such a mapping from the unit square onto the parametric domain that the corresponding mapping of the surface $f(u, v)$ onto the parametric domain is harmonic. Further theoretical details can be found in [39].

Let us introduce the following data structure suitable for either regular or block-structured grids. $C_O(N, k)$ defines correspondence between the local and the global node numbers.

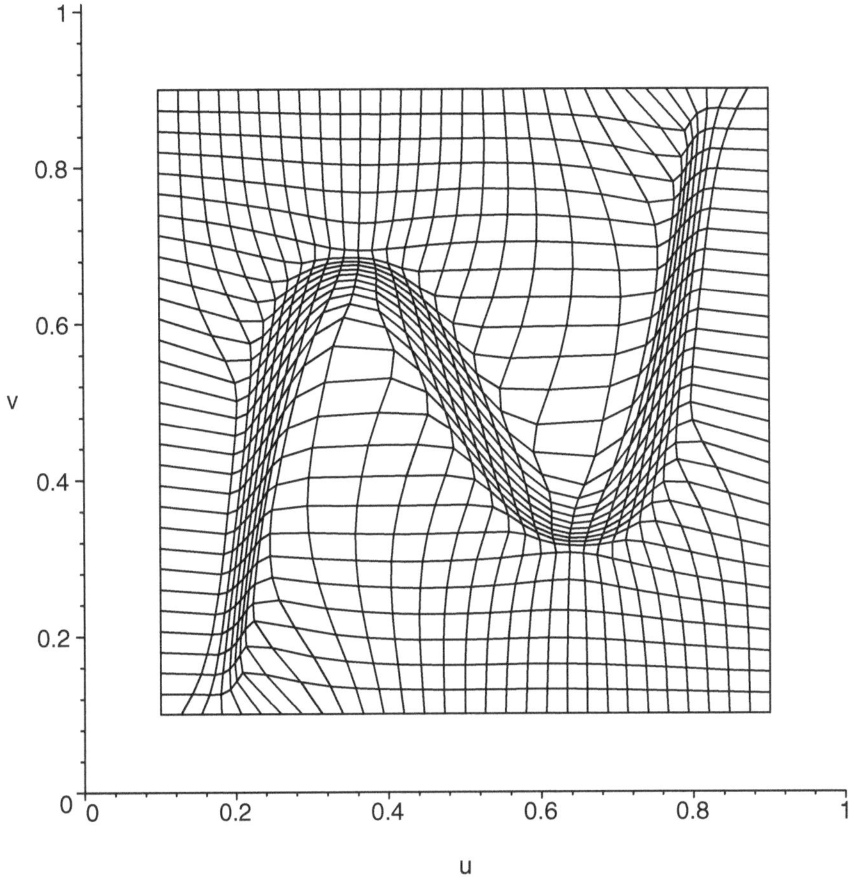

(a)

Fig. 5.14a. Example of adaptive-harmonic grid

$$C_O(N,k) = n \quad n = 1,\ldots,N_n \quad N = 1,\ldots,N_e \quad k = 1,2,3,4, \tag{5.14}$$

where n is a global node number, N_n is a total number of grid nodes, N is an element number, N_e is a number of elements, k is a local node number in the element. The cell vertices are numbered from 1 to 4 in the clockwise direction. Each vertex is associated with a triangle: vertex 1 with $\triangle_{412}$, vertex 2 with $\triangle_{123}$, vertex 3 with $\triangle_{234}$, and vertex 4 with $\triangle_{341}$. It can be easily demonstrated that $2A_{k-1,k,k+1} = J_k$ [39], where A denotes the area of the corresponding triangle and $J_k, k = 1,2,3,4$ is the Jacobian of the mapping given by

$$J_k = (u_{k-1} - u_k)(v_{k+1} - v_k) - (u_{k+1} - u_k)(v_{k-1} - v_k), \tag{5.15}$$

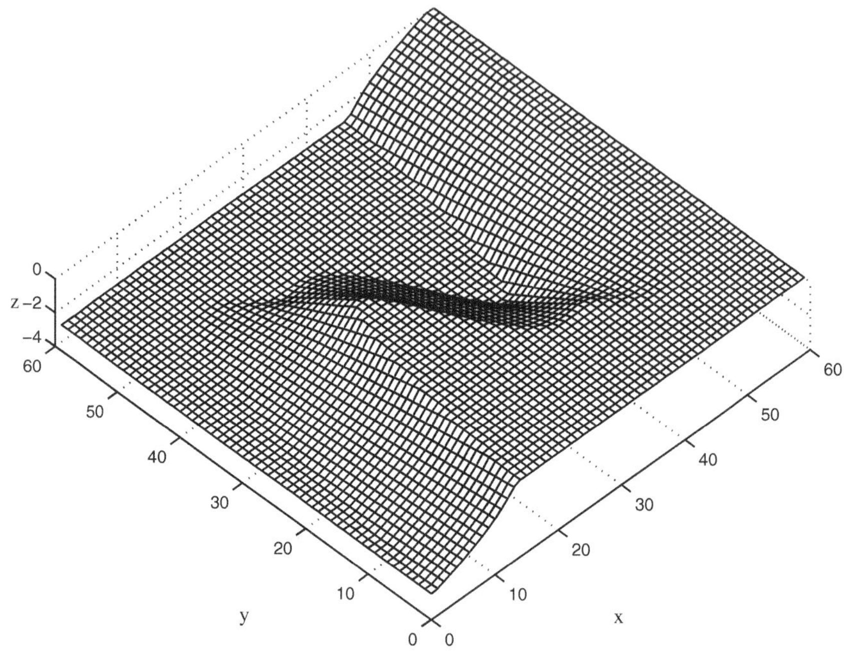

(b)

Fig. 5.14b. Control function

where $k-1=4$ if $k=1$ and $k+1=1$ if $k=4$. An example of such data structure is shown in Fig. 5.15.

The function $N'(n,d)$ defines the correspondence between the grid node n, and its neighbors in the d direction, $d \in$ {left,right,up,down}. Functional (5.12) is approximated by a discrete functional as follows

$$I \approx I^h = \sum_{N=1}^{N_e} \sum_{k=1}^{4} \frac{1}{4} [F_k]_N. \tag{5.16}$$

where

$$\begin{aligned} F_k &= \frac{D_1\left[1+(f_u)_k^2\right] + D_2\left[1+(f_v)_k^2\right] + 2D_3 (f_u)_k (f_v)_k}{J_k\left[1+(f_u)_k^2+(f_v)_k^2\right]^{\frac{1}{2}}}, \\ D_1 &= (u_{k-1}-u_k)^2 + (u_{k+1}-u_k)^2, \\ D_2 &= (v_{k-1}-v_k)^2 + (v_{k+1}-v_k)^2, \\ D_3 &= (u_{k-1}-u_k)(v_{k-1}-v_k) + (u_{k+1}-u_k)(v_{k+1}-v_k), \\ J_k &= (u_{k-1}-u_k)(v_{k+1}-v_k) - (u_{k+1}-u_k)(v_{k-1}-v_k). \end{aligned} \tag{5.17}$$

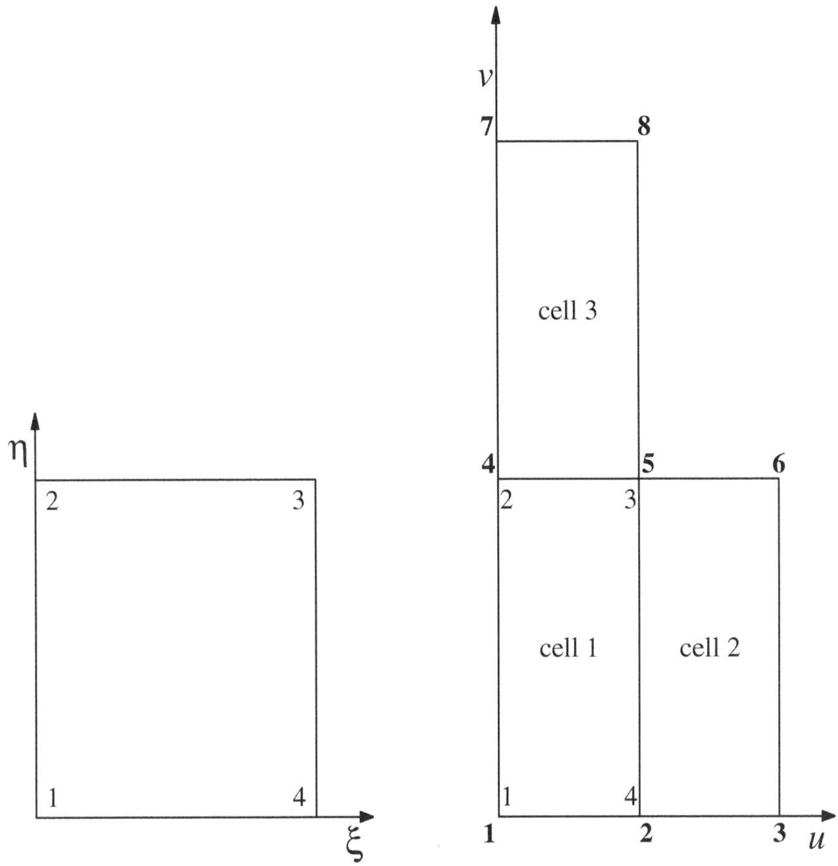

Fig. 5.15. Correspondence of node numbers for a mapping of the unit square in the (ξ, η) plane on to the quadrilateral cell 1 of the grid in the (u, v) plane

Here $(f_u)_k$ and $(f_v)_k$ are the derivatives of the control function in the u and v direction, respectively, computed in the node number k of the cell number N. The convexity of the grid cells requires that the Jacobian be positive [37], i.e.,

$$[J_k]_N > 0 \quad k = 1, 2, 3, 4 \quad N = 1, \ldots, N_e, \tag{5.18}$$

The discrete functional (5.16) is minimized numerically using an initial grid as the first iteration. The minimization is arranged in such a way that the minimum is attained on a grid composed of convex quadrilaterals. This property improves the stability of the grid generation procedure. Moreover, it can be shown that if the set of convex grids corresponding to the mapping is not empty, the system of algebraic equations

$$R_u = \frac{\partial I^h}{\partial u_n} = 0,$$
$$R_v = \frac{\partial I^h}{\partial v_n} = 0, \tag{5.19}$$

has at least one solution which is a convex grid.

Suppose the grid at the lth step of the iteration is determined, the grid at the $(l+1)^{\text{th}}$ step is computed using the quasi-Newtonian procedure as follows:

$$u_n^{l+1} = u_n^l - \tau\left(R_u\frac{\partial R_v}{\partial v_n} - R_v\frac{\partial R_u}{\partial v_n}\right)\left(\frac{\partial R_u}{\partial u_n}\frac{\partial R_v}{\partial v_n} - \frac{\partial R_v}{\partial u_n}\frac{\partial R_u}{\partial v_n}\right)^{-1},$$
$$v_n^{l+1} = v_n^l - \tau\left(R_v\frac{\partial R_u}{\partial u_n} - R_u\frac{\partial R_v}{\partial u_n}\right)\left(\frac{\partial R_u}{\partial u_n}\frac{\partial R_v}{\partial v_n} - \frac{\partial R_v}{\partial u_n}\frac{\partial R_u}{\partial v_n}\right)^{-1}, \tag{5.20}$$

where τ is the iteration parameter, which is chosen so that the grid remains convex. At each step, condition (5.18) is verified. If (5.18) is not satisfied, then τ is reduced by half and the grid at the $(l+1)^{\text{th}}$ step is re-computed. The derivation of the computational formulas for (5.20) suitable for computer programming is given in the Appendix.

In the case of the tool path generation, the gradient of the control function f corresponds to the machining error (scallop) between the tool tracks. Since the tool path is a discrete set of points the derivatives of the control function $f_u(u,v)$ and $f_v(u,v)$ for a given surface can not be explicitly evaluated. Therefore, these derivatives are generated "artificially" as follows. Initially, $(f_u)_n$ and $(f_v)_n$ are set to 0. Next,

$$(f_u)_n^{l+1} = \begin{cases}(f_u)_n^l + \lambda_+ & \text{if } s_u(n) > 0,\\ (f_u)_n^l - \lambda_- & \text{otherwise,}\end{cases}$$
$$(f_v)_n^{l+1} = \begin{cases}(f_v)_n^l + \lambda_+ & \text{if } s_v(n) > 0,\\ (f_v)_n^l - \lambda_- & \text{otherwise,}\end{cases} \tag{5.21}$$

where λ_+ and λ_- are the prescribed increment and decrement, respectively, and $s_u(n)$ and $s_v(n)$ are the scallop estimates evaluated at the node n as follows:

$$s_u(n) = \max_{d\in\{\text{left,right}\}}\left[D(n,N'(n,d)) - T(n,N'(n,d))\right],$$
$$s_v(n) = \max_{d\in\{\text{up,down}\}}\left[D(n,N'(n,d)) - T(n,N'(n,d))\right]. \tag{5.22}$$

where $D(n,m)$ is the distance between nodes n and m given by

$$D(n,m) = |S(u_n,v_n) - S(u_m,v_m)|. \tag{5.23}$$

$T(n,m)$ is an estimate of the machining strip width calculated at the midpoint $S(\frac{u_n+u_m}{2}, \frac{v_n+v_m}{2})$.

The grid generation algorithm consists of the following steps:

1. Generate the initial convex grid. The grid is generated manually or by interpolating. Note that interpolation may generate a grid with the nodes outside the boundary of the region. In that case, several iterations can be performed by the classic Brackbill's and Saltzman's method [15] which will move the nodes back inside the region. The initial grid should not satisfy the scallop constraints. If the grid satisfies the constraints for every node then the adaptation is not necessary. On the other hand, it means that the number of nodes has been overestimated and should be reduced. The initial grid size (l_r rows and l_c columns) can be estimated as follows

$$\begin{aligned} l_r &= \left\lceil \frac{W_v}{2r} \right\rceil, \\ l_c &= \left\lceil \frac{W_u}{2r} \right\rceil, \end{aligned} \tag{5.24}$$

where W_u and W_v are the surface dimensions in the u and v directions, respectively and r is the radius of the cutting tool.

2. Adapt the grid by minimizing the Dirichlet functional until a prescribed number of iterations has been performed.

3. Refine the grid by insering additional nodes. If the grid does not satisfy the scallop height constraint, more grid points are needed. For a structured grid having l_r rows and l_c columns, the numbers of rows and columns at the next step are estimated as follows

$$\begin{aligned} l_r^{\text{new}} &= l_r + \left\lceil \frac{\max_{1\le i\le l_r}\left(\sum_{j=1}^{l_c-1}\left[D\left(n_{i,j}, n_{i,j+1}\right) - T\left(n_{i,j}, n_{i,j+1}\right)\right]\right)}{2r} \right\rceil \\ l_c^{\text{new}} &= l_c + \left\lceil \frac{\max_{1\le j\le l_c}\left(\sum_{i=1}^{l_r-1}\left[D\left(n_{i,j}, n_{i+1,j}\right) - T\left(n_{i,j}, n_{i+1,j}\right)\right]\right)}{2r} \right\rceil, \end{aligned} \tag{5.25}$$

where $n_{i,j}$ is the grid node on row i and column j. For block-structured grid, the grid is first partitioned into smaller blocks of structured grids and the refinement is performed for each block. For column insertion, the grid is partitioned vertically as shown in Fig. 5.16(b) whereas for row insertion, the grid is partitioned horizontally as shown in Fig. 5.16(c) (see also Fig. 5.23 in Example 5.6).

4. Adapt the grid by minimizing the Dirichlet functional until all the grid points satisfy the scallop constraint or until a prescribed number of iterations has been performed.

5. If the scallop constraint has not been satisfied for some points, goto the refinement step 3.

The algorithm converges faster and exhibits more stability if the third term

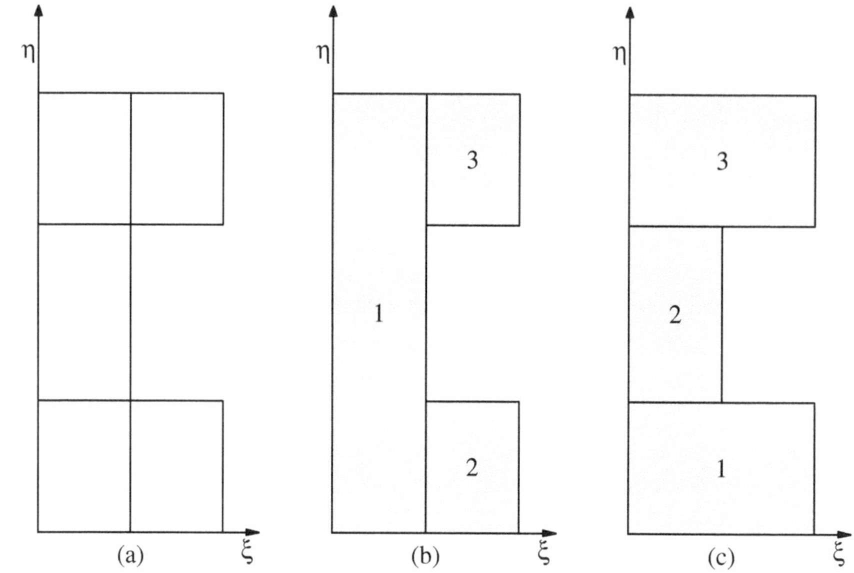

Fig. 5.16. Partitioning of block-structured grid for refinement

in the nominator of functional (5.12) is neglected. In order to interpret this term geometrically for each F_k in (5.16), consider two vectors from local node k to local nodes $k-1$ and $k+1$ given by

$$\begin{aligned} \mathbf{v}_{k,k-1} &= [u_k - u_{k-1}, v_k - v_{k-1}], \\ \mathbf{v}_{k,k+1} &= [u_k - u_{k+1}, v_k - v_{k+1}]. \end{aligned} \tag{5.26}$$

Clearly, $\mathbf{v}_{k,k-1} \cdot \mathbf{v}_{k,k+1} = (u_k - u_{k-1})(u_k - u_{k+1}) + (v_k - v_{k-1})(v_k - v_{k+1}) = D_3 = |\mathbf{v}_{k,k-1}||\mathbf{v}_{k,k+1}| \cos\vartheta$, where ϑ is the angle between the two vectors. Therefore

$$F_k = \frac{\frac{D_1}{c}\left[1 + (f_u)_k^2\right] + \frac{D_2}{c}\left[1 + (f_v)_k^2\right] + 2(f_u)_k(f_v)_k \cos\vartheta}{\frac{J_k}{c}\left[1 + (f_u)_k^2 + (f_v)_k^2\right]^{\frac{1}{2}}}, \tag{5.27}$$

where $c = |\mathbf{v}_{k,k-1}||\mathbf{v}_{k,k+1}|$. Therefore, omitting the last term leads to an adaptation with less impact on the angles between coordinate lines. Finally, recall that convexity of the cells is provided by an appropriate τ in (6).

5.5 Space-Filling Curve Generation on Block Structured Grid

Space-filling curve tool paths on block-structured grids can be generated by imposing one extra requirement on the grid generation algorithm presented in the previous section. At each refinement stage, the size of any sub-block of the grid partitioned vertically or horizontally must be even. This is a sufficient condition for the existing of Hamiltonian path, i.e., the space-filling curve on the block structured grid. The proof of this claim is quite straightforward since all nodes in the blocks of the even size can all be covered by small rectangular circuits. The merging stage of the space-filling curve generation algorithm then merges all small circuits into one bigger circuit. Figure 5.17 illustrates the space-filling curve generation on block structured grid. First, a vertex is placed at the center of each grid cell (Fig. 5.17(b)). Each vertex is then connected to the vertices of the neighbor cells (Fig. 5.17(c)). Finally, small circuits are created by constructing small rectangular cyclic paths over every 4 adjacent vertices, i.e, by connecting the vertices on even rows and columns with the vertices on odd rows and columns, respectively, as shown in Fig. 5.18. Merging of small circuits to create the space-filling curve is then performed (see Sect. 4.3.2).

The combination of space-filling curve and the grid generation techniques allows construction of tool paths on surfaces with complex irregular boundaries, cuts off, pockets, islands, etc. Examples are given in the next section.

5.6 Examples and Discussion

This section presents a variety of examples that demonstrate the efficiency and advantages of the use of curvilinear grid in tool path generation.

Example 5.5. This example demonstrates the use of variational grid generation technique to overcome the inefficiency of tool path generation by the conventional isoparametric method. Consider a unimodal surface with an exponential peak along a line in the parametric domain (u, v). The surface is given by (see Fig. 5.19)

$$\begin{aligned} x &= 100u - 50, \\ y &= 100v - 50, \\ z &= 10\exp(-40(2u - 0.5 - v)^2) - 15. \end{aligned} \tag{5.28}$$

A curvilinear grid constructed for a flat-end tool with radius 3 mm and the surface tolerance of 0.1 mm is shown in Fig. 5.20. The comparison of the zigzag and SFC tool paths generated by traditional isoparametric method and curvilinear grid method is presented in Table 5.4. The zigzag tool path based on the adaptive grid is shown in Fig. 5.21(c). The length of the tool path is 2558.28 mm. The SFC tool path requires a basic (isoparametric) grid with a small spatial step. Consequently it leads to a SFC of about 3066.87 mm (Fig. 5.21(b)). Finally, the curvilinear grid adapted to the zones where the small distance between the tracks is required leads to a shorter path of about 2519.88 mm (Fig. 5.21(d)). The length of the zigzag and SFC tool paths

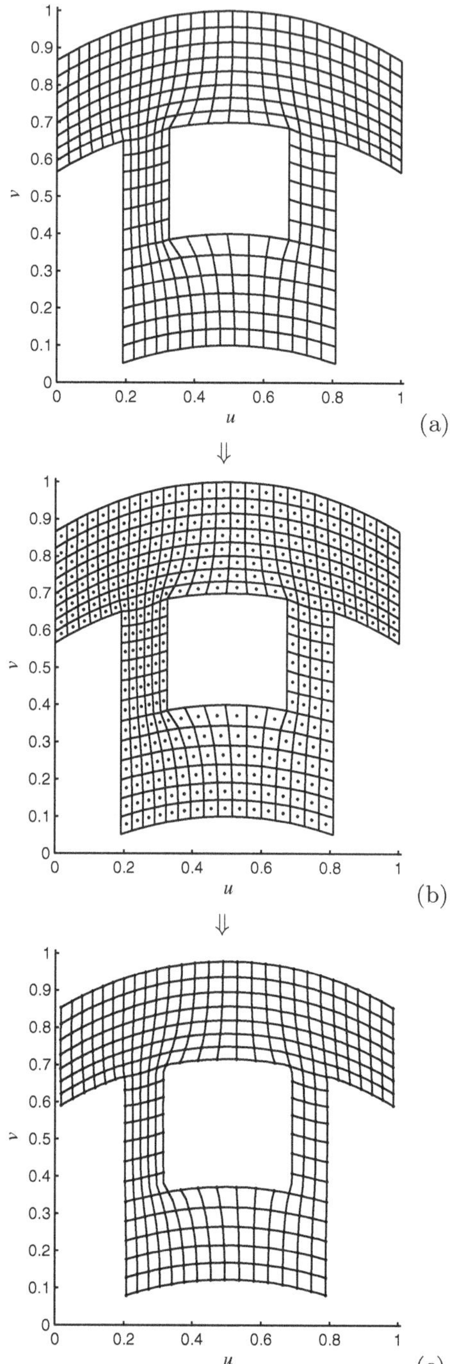

Fig. 5.17. Undirected graph construction for SFC tool path generation

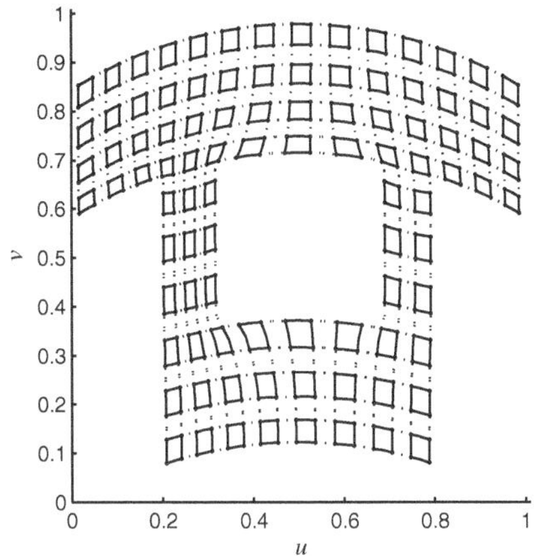

Fig. 5.18. Covering of grid in Fig. 5.17 by small circuits

based on the adaptive grid are shorter by 45.76% and 17.84%, respectively, when compared with the zigzag and SFC based on isoparametric tool path method (Fig. 5.21).

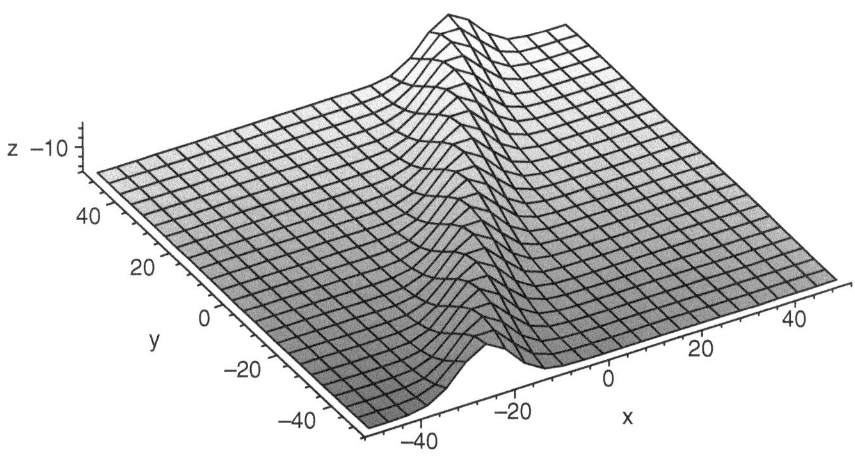

Fig. 5.19. Unimodal surface with exponential peak along a line

The SFC tool path is better than the standard zigzag path since it turns in the optimal direction, that is, in the direction which maximizes the machining strip. Furthermore, a step between two adjusted tracks on the rectangular grid is set to the minimal step ensuring the required scallop along the two tracks. The curvilinear grid makes it possible to vary the distance between the tracks making the step small only where necessary. Therefore, the SFC constructed within a new basic grid leads to better results.

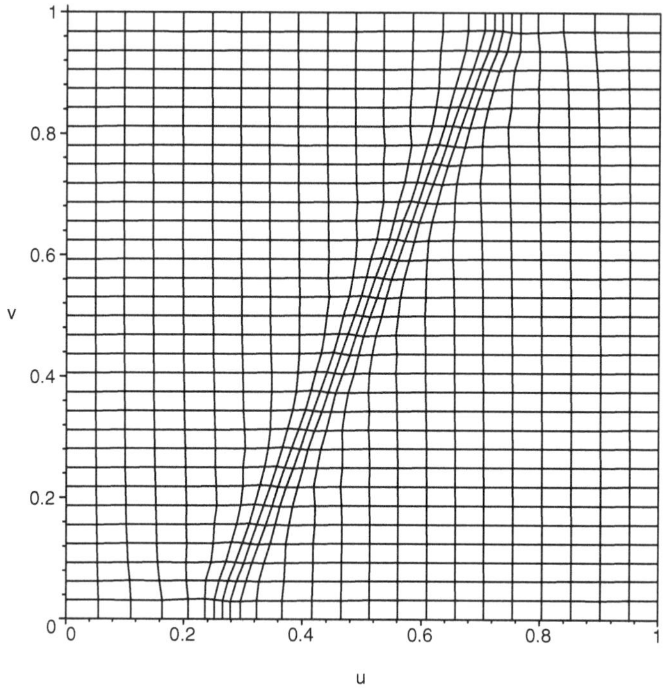

(a)

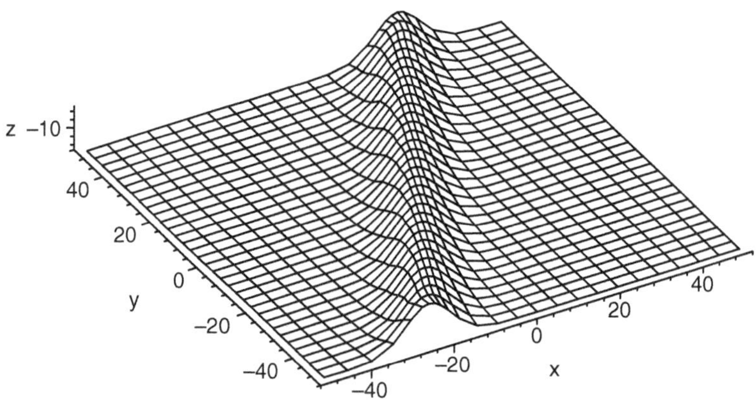

(b)

Fig. 5.20. Curvilinear grid adapted to the unimodal surface with exponential peak along a line in (u, v) domain (**a**) and workpiece coordinate system (**b**)

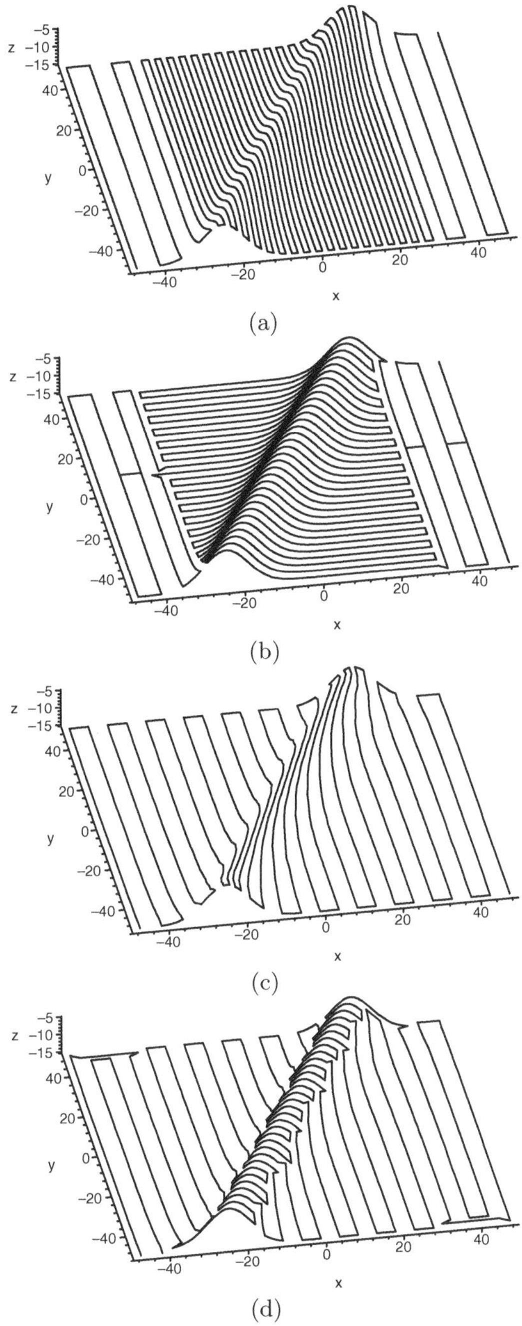

Fig. 5.21. (**a**) isoparametric tool path, (**b**) SFC tool path from two isoparametric tool paths, (**c**) zigzag tool path from adaptive grid, (**d**) SFC tool path from adaptive grid

Table 5.4. Comparison of tool paths based on variational grid generation techniques versus the isoparametric tool path scheme

	Total path length (mm)	
Method used for constructing grid	Vertical zigzag	SFC
Overlaying of two isoparametric tool paths	4716.91	3066.87
Curvilinear grid generation	2557.74	2508.25

The plots of convergence rates are shown in Fig. 5.22. Note that the convergence rate of the algorithm increases a little when the third term of the numerator in Direchlet functional (5.12) is dropped. Although, the improvement in convergence rate is not evidently large, the complexity of the algorithm is reduced by eliminating this third term. Consequently, the grid requires less amount of time to compute.

Example 5.6. This example demonstrates the use of curvilinear grid and SFC to construct tool paths to machine surfaces with complex irregular boundaries, cuts off, and islands. Consider a two-bell surface given by (4.8) in Example 4.3. The surface is to be machined only in the area within the boundaries shown in Fig. 5.24(a) in (u, v) domain. The initial grid and the first refined grid are depicted in Fig. 5.23. The final grid and SFC tool path are shown in Figs. 5.24 and 5.25, respectively. The simulaton of the cutting in Unigraphics 18 for a flat-end tool with radius 3 mm and the surface tolerance of 0.05 mm is shown in Fig. 5.26. The plots of convergence rates are shown in Fig. 5.27. The convergence rate of the algorithm also increases a little when the third term of the numerator in Direchlet functional (5.12) is dropped.

Example 5.7. Consider a single blade of an impeller depicted in Fig. 5.28(a). This single blade is described by parametric equations. When the blade is broken (see Fig. 5.28(b)), it is sometimes possible to fill the broken blade with a new material and mill only the filled part. Since this part of the blade contains irregular boundaries in (u, v) domain, the traditional isoparametric method cannot be used to construct the tool path. Applying the grid generation method on this parametric surface with complex irregular boundaries, the final curvilinear grid adapted to the shape of the broken blade is shown in Fig. 5.29 (parametric domain) and Fig. 5.30 (workpiece coordinates). The grid is adapted to the surface that is to be milled using ball-end tool of radius 5 mm and machined surface tolerance of 0.05 mm. The calculation of machining strip width for ball-end mill is given in Sect. 3.2. The SFC tool path is shown in Fig. 5.31 (parametric domain) and Fig. 5.32 (workpiece coordinates). The virtual cutting with the SFC tool path is shown in Fig. 5.33.

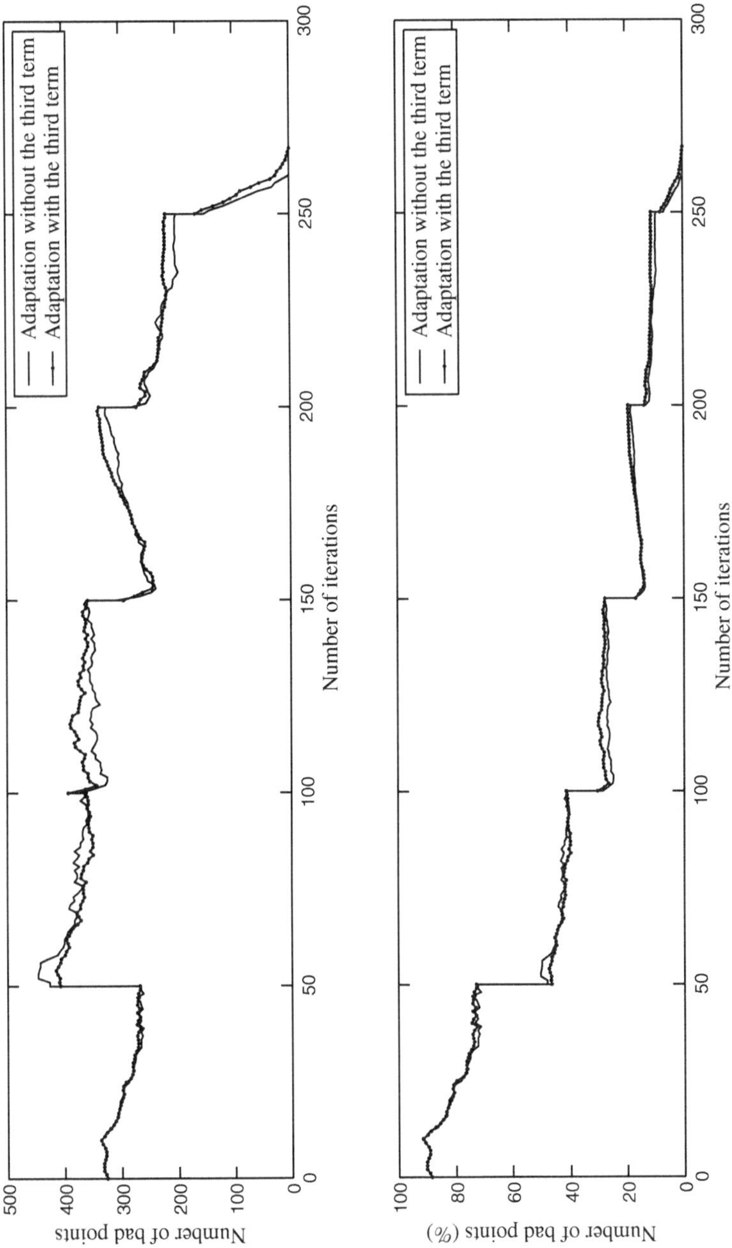

Fig. 5.22. Convergence rate of the grid generation technique in term of the number of iterations for Example 5.4

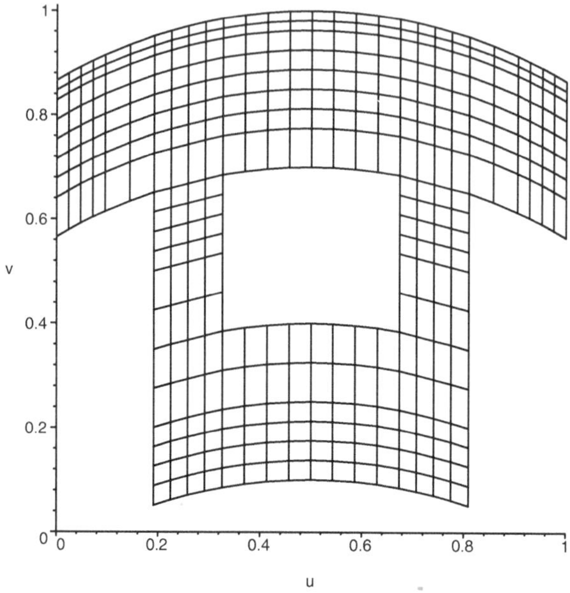

(a)

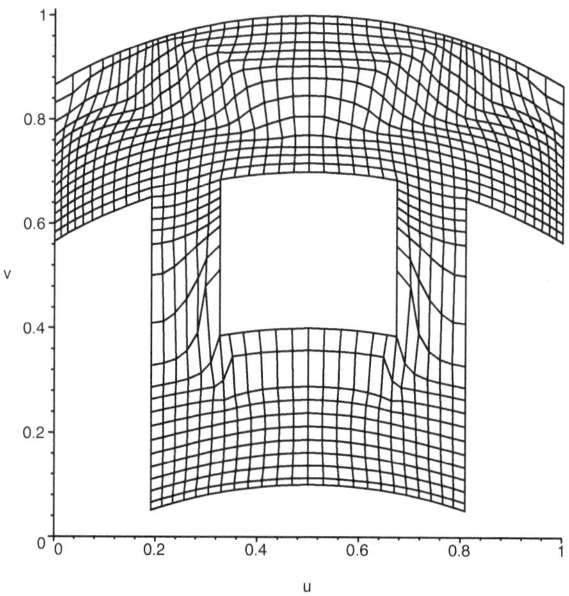

(b)

Fig. 5.23. Grid refinement: (**a**) initial grid, (**b**) grid after first refinement

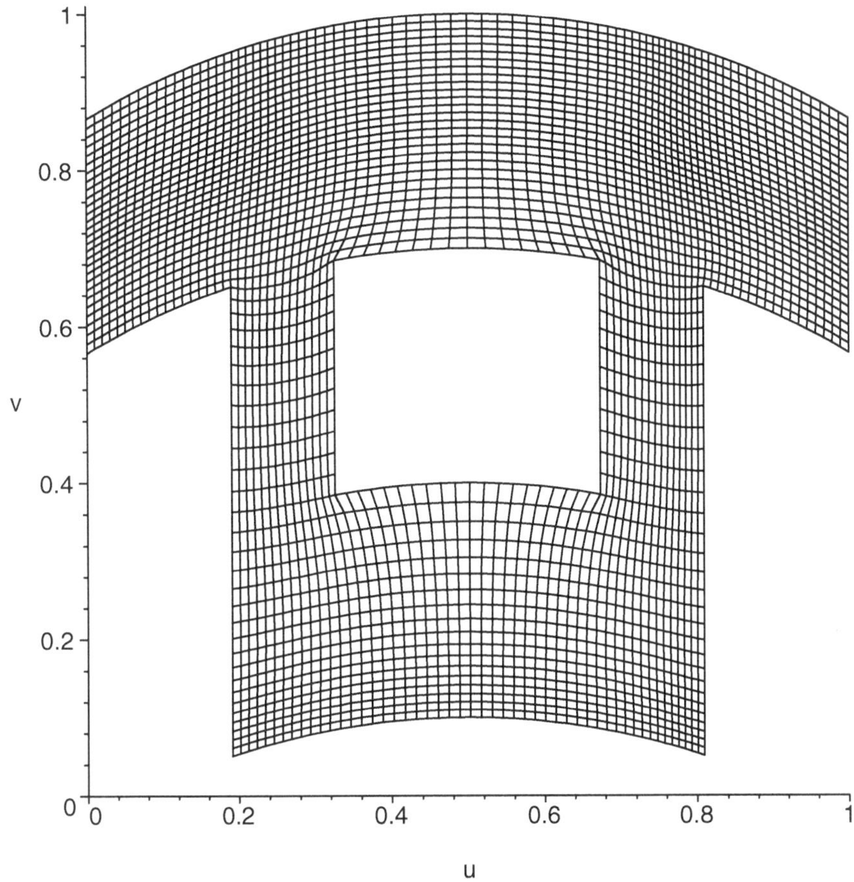

Fig. 5.24a. Curvilinear grid adapted to the surface in Example 5.6 in (u, v) domain

Appendix: Derivation of Computational Formulas for Adaptive-Harmonic Grid Generation

The minimization of the discrete functional (5.16) requires computation of F_k and its derivatives. Each computation of F_k involves three grid nodes on the triangle $\triangle_{k-1,k,k+1}$. For conciseness, node $k-1$, k and $k+1$ are denoted by a, b and c, respectively. F_k is denoted by F and is now expressed as

$$F = \frac{X}{Y}, \tag{5.29}$$

where

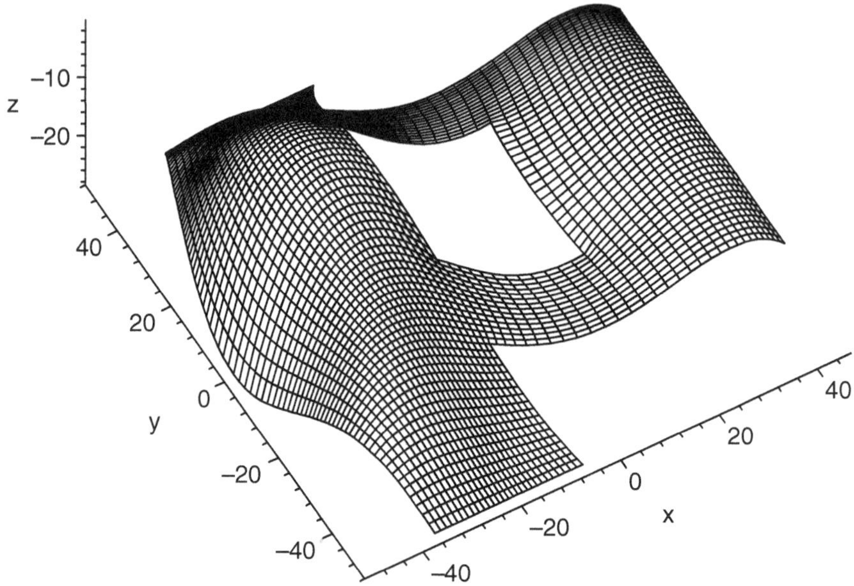

Fig. 5.24b. Curvilinear grid adapted to the surface in Example 5.6 in workpiece coordinate system

$$
\begin{aligned}
X &= \frac{D_1\left[1+(f_u)_k^2\right]+D_2\left[1+(f_v)_k^2\right]+2D_3(f_u)_k(f_v)_k}{\left[1+(f_u)_k^2+(f_v)_k^2\right]^{\frac{1}{2}}},\\
Y &= (u_a-u_b)(v_c-v_b)-(u_c-u_b)(v_a-v_b),\\
D_1 &= (u_a-u_b)^2+(u_c-u_b)^2,\\
D_2 &= (v_a-v_b)^2+(v_c-v_b)^2,\\
D_3 &= (u_a-u_b)(v_a-v_b)+(u_c-u_b)(v_c-v_b).
\end{aligned}
\tag{5.30}
$$

The formulas for the derivatives of F can then be expressed as the derivatives of two functions X and Y and are given below

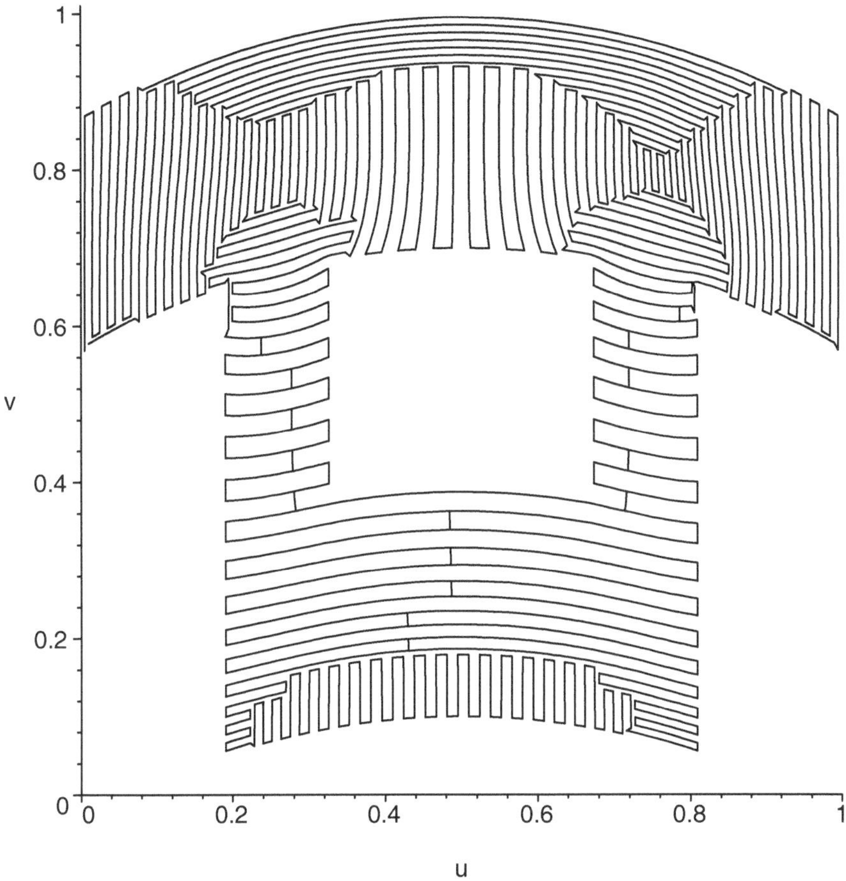

Fig. 5.25a. SFC tool path for the surface in Example 5.6 in (u, v) domain

$$
\begin{aligned}
F_u &= \frac{X_u Y - X Y_u}{Y^2} = \frac{X_u - F Y_u}{Y}, \\
F_v &= \frac{X_v Y - X Y_v}{Y^2} = \frac{X_v - F Y_v}{Y}, \\
F_{uu} &= \frac{(X_{uu} - F_u Y_u - F Y_{uu})Y - (X_u - F Y_u)Y_u}{Y^2}, \\
&= \frac{X_{uu} - 2F_u Y_u - F Y_{uu}}{Y}, \\
F_{vv} &= \frac{(X_{vv} - F_v Y_v - F Y_{vv})Y - (X_v - F Y_v)Y_v}{Y^2}, \\
&= \frac{X_{vv} - 2F_v Y_v - F Y_{vv}}{Y}, \\
F_{uv} &= \frac{(X_{uv} - F_v Y_u - F Y_{uv})Y - (X_u - F Y_u)Y_v}{Y^2}, \\
&= \frac{X_{uv} - F_v Y_u - F_u Y_v - F Y_{uv}}{Y}.
\end{aligned}
\tag{5.31}
$$

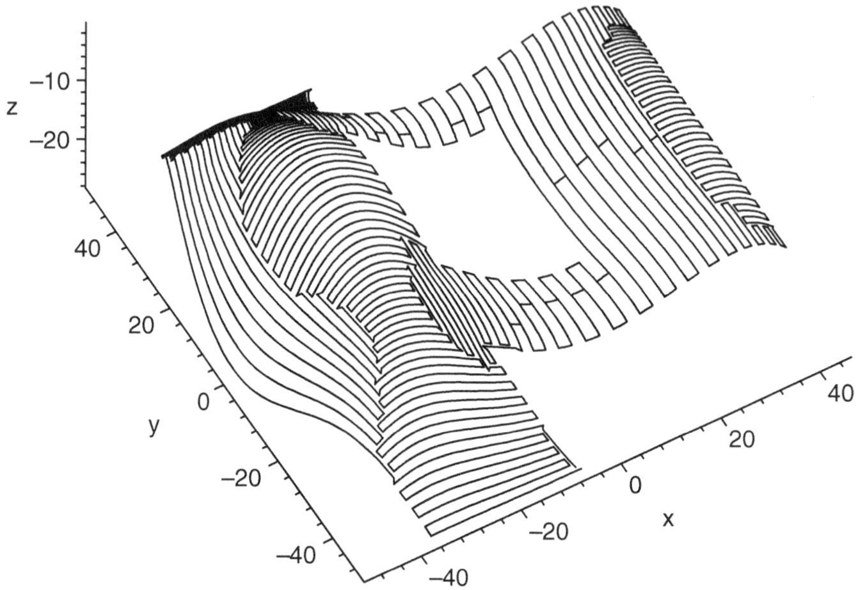

Fig. 5.25b. SFC tool path for the surface in Example 5.6 in workpiece coordinate system

The derivatives at node a are obtained by substituting u by u_a and v by v_a and are given by

$$
\begin{aligned}
&Y_u = v_c - v_b, \quad Y_v = u_b - u_c, \quad Y_{uu} = 0, \quad Y_{vv} = 0, \quad Y_{uv} = 0,\\
&X_u = 2\frac{\left[1+(f_u)_k^2\right](u_a - u_b) + (f_u)_k(f_v)_k(v_a - v_b)}{\left[1+(f_u)_k^2+(f_v)_k^2\right]^{\frac{1}{2}}},\\
&X_v = 2\frac{\left[1+(f_v)_k^2\right](v_a - v_b) + (f_u)_k(f_v)_k(u_a - u_b)}{\left[1+(f_u)_k^2+(f_v)_k^2\right]^{\frac{1}{2}}},\\
&X_{uu} = 2\frac{1+(f_u)_k^2}{\left[1+(f_u)_k^2+(f_v)_k^2\right]^{\frac{1}{2}}}, \quad X_{vv} = 2\frac{1+(f_v)_k^2}{\left[1+(f_u)_k^2+(f_v)_k^2\right]^{\frac{1}{2}}},\\
&X_{uv} = 2\frac{(f_u)_k(f_v)_k}{\left[1+(f_u)_k^2+(f_v)_k^2\right]^{\frac{1}{2}}}.
\end{aligned}
\tag{5.32}
$$

The derivatives at node b are obtained by substituting u by u_b and v by v_b and are given by

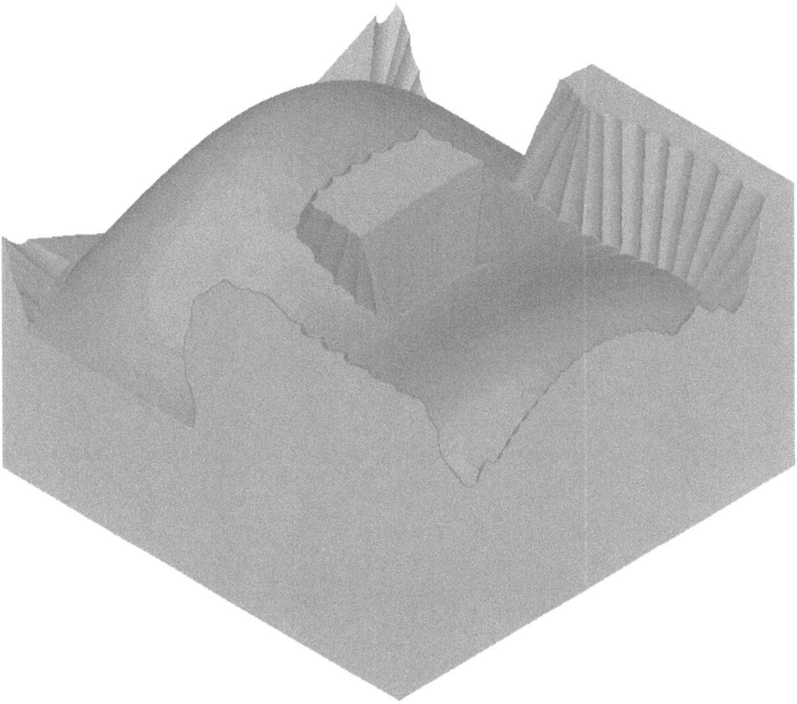

Fig. 5.26. Simulation result of five-axis machining with SFC tool path in Unigraphics 18 for the surface in Example 5.6

$$
\begin{aligned}
&Y_u = v_a - v_c, \quad Y_v = u_c - u_a, \quad Y_{uu} = 0, \quad Y_{vv} = 0, \quad Y_{uv} = 0, \\
&X_u = 2\frac{\left[1 + (f_u)_k^2\right](2u_b - u_a - u_c) + (f_u)_k(f_v)_k(2v_b - v_a - v_c)}{[1 + (f_u)_k^2 + (f_v)_k^2]^{\frac{1}{2}}}, \\
&X_v = 2\frac{\left[1 + (f_v)_k^2\right](2v_b - v_a - v_c) + (f_u)_k(f_v)_k(2u_b - u_a - u_c)}{[1 + (f_u)_k^2 + (f_v)_k^2]^{\frac{1}{2}}}, \\
&X_{uu} = 4\frac{1 + (f_u)_k^2}{[1 + (f_u)_k^2 + (f_v)_k^2]^{\frac{1}{2}}}, \quad X_{vv} = 4\frac{1 + (f_v)_k^2}{[1 + (f_u)_k^2 + (f_v)_k^2]^{\frac{1}{2}}}, \\
&X_{uv} = 4\frac{(f_u)_k(f_v)_k}{[1 + (f_u)_k^2 + (f_v)_k^2]^{\frac{1}{2}}}.
\end{aligned}
\tag{5.33}
$$

Finally, the derivatives at node c are obtained by substituting u by u_c and v by v_c and are given by

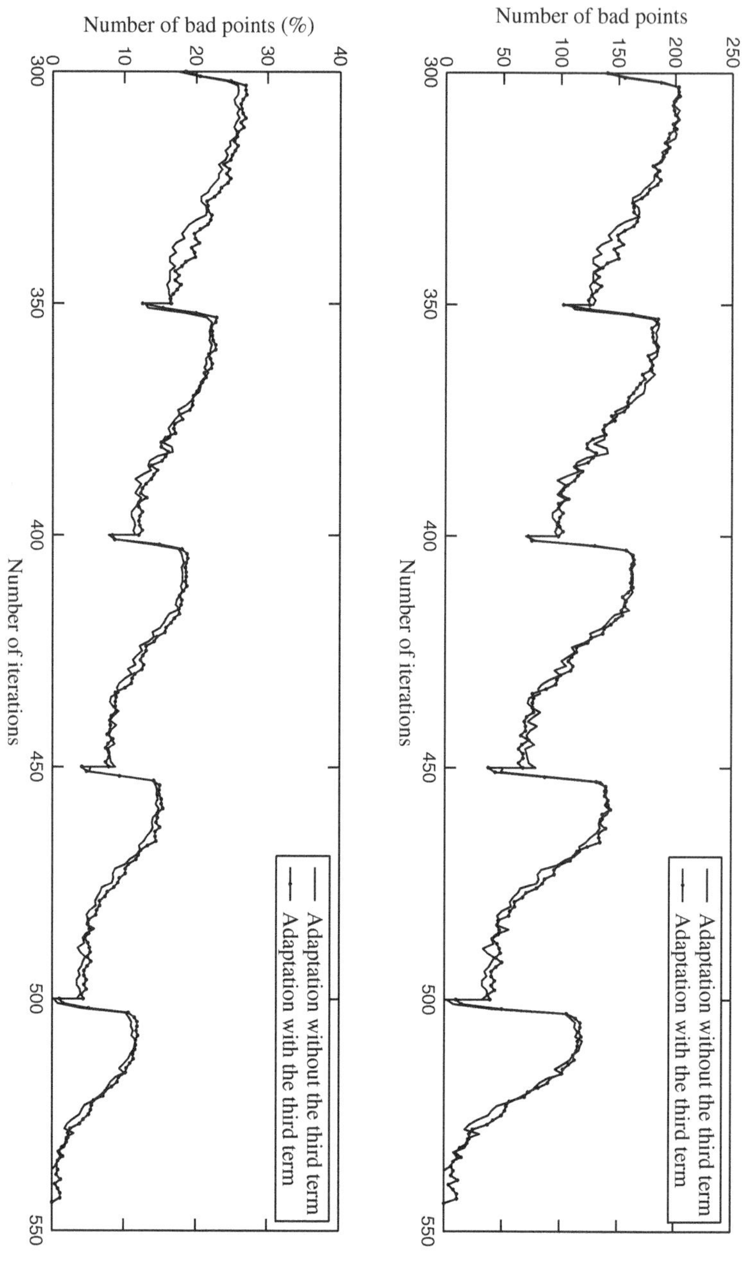

Fig. 5.27. Convergence rate of the grid generation technique in term of the number of iterations for Example 5.6

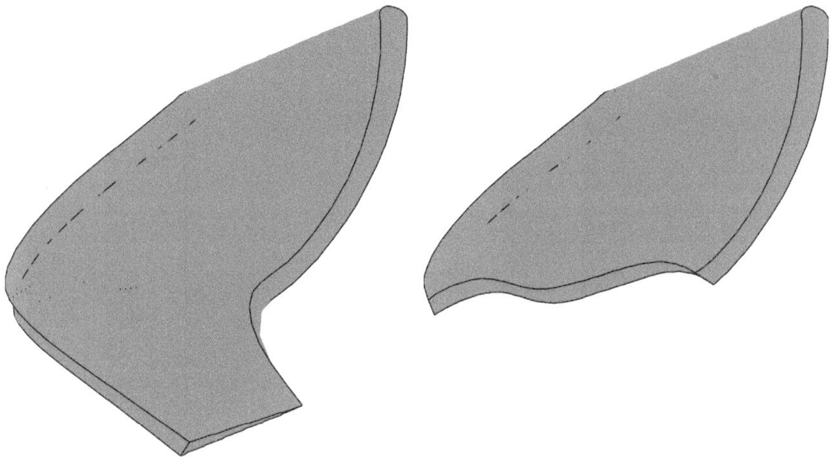

(a) (b)

Fig. 5.28. (**a**) a full single blade of an impeller, (**b**) part of a single blade

$$
\begin{gathered}
Y_u = v_b - v_a, \quad Y_v = u_a - u_b, \quad Y_{uu} = 0, \quad Y_{vv} = 0, \quad Y_{uv} = 0, \\
X_u = 2\frac{\left[1+(f_u)_k^2\right](u_c - u_b) + (f_u)_k (f_v)_k (v_c - v_b)}{\left[1+(f_u)_k^2 + (f_v)_k^2\right]^{\frac{1}{2}}}, \\
X_v = 2\frac{\left[1+(f_v)_k^2\right](v_c - v_b) + (f_u)_k (f_v)_k (u_c - u_b)}{\left[1+(f_u)_k^2 + (f_v)_k^2\right]^{\frac{1}{2}}}, \\
X_{uu} = 2\frac{1+(f_u)_k^2}{\left[1+(f_u)_k^2 + (f_v)_k^2\right]^{\frac{1}{2}}}, \quad X_{vv} = 2\frac{1+(f_v)_k^2}{\left[1+(f_u)_k^2 + (f_v)_k^2\right]^{\frac{1}{2}}}, \\
X_{uv} = 2\frac{(f_u)_k (f_v)_k}{\left[1+(f_u)_k^2 + (f_v)_k^2\right]^{\frac{1}{2}}}.
\end{gathered}
\tag{5.34}
$$

For the cell number N, $N \in \{1, \ldots, N_e\}$ and triangle number k, $k \in \{1, 2, 3, 4\}$, the values of F and its derivatives on u_a and v_a are computed with the use of (5.31) and (5.32). The computed values are then added to the appropriate array elements

$$
\begin{aligned}
I^h_{\text{new}} &= I^h_{\text{old}} + F, \\
[R_u]_{n,\text{new}} &= [R_u]_{n,\text{old}} + F_u, \\
[R_v]_{n,\text{new}} &= [R_v]_{n,\text{old}} + F_v, \\
[R_{uu}]_{n,\text{new}} &= [R_{uu}]_{n,\text{old}} + F_{uu}, \\
[R_{vv}]_{n,\text{new}} &= [R_{vv}]_{n,\text{old}} + F_{vv}, \\
[R_{uv}]_{n,\text{new}} &= [R_{uv}]_{n,\text{old}} + F_{uv},
\end{aligned}
\tag{5.35}
$$

where $n = C_O(N, k-1)$.

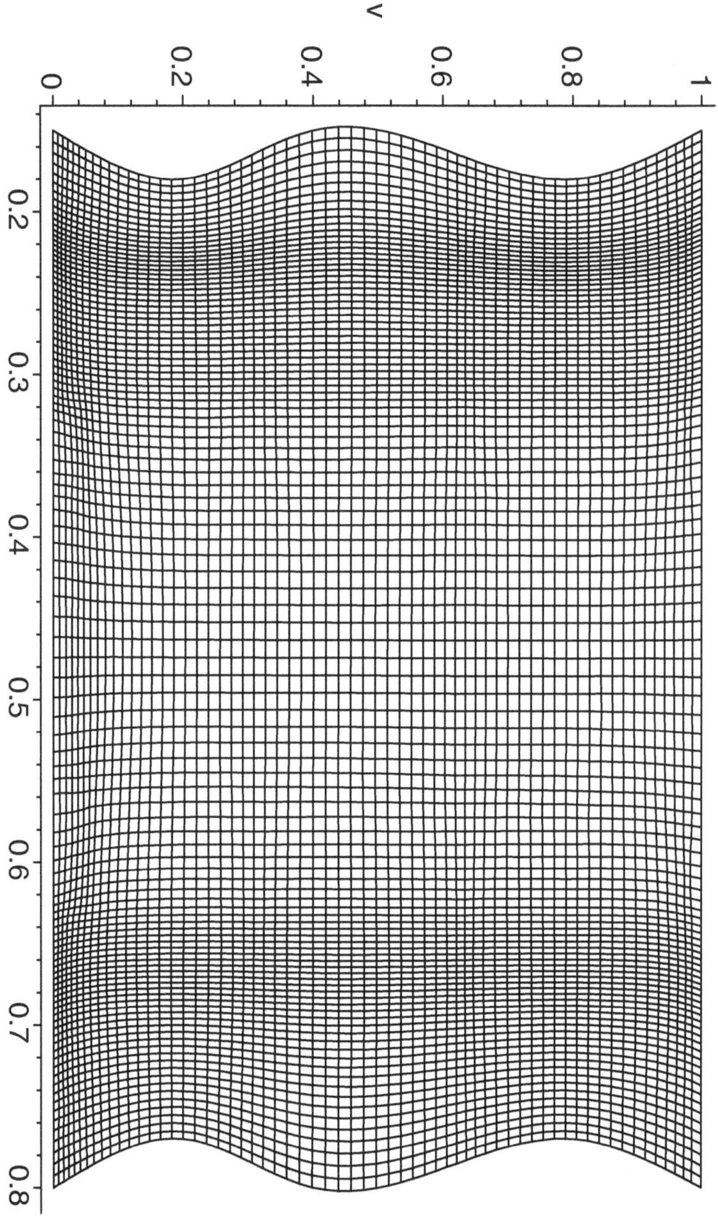

Fig. 5.29. Curvilinear grid adapted to part of a surface of the blade in (u, v) domain

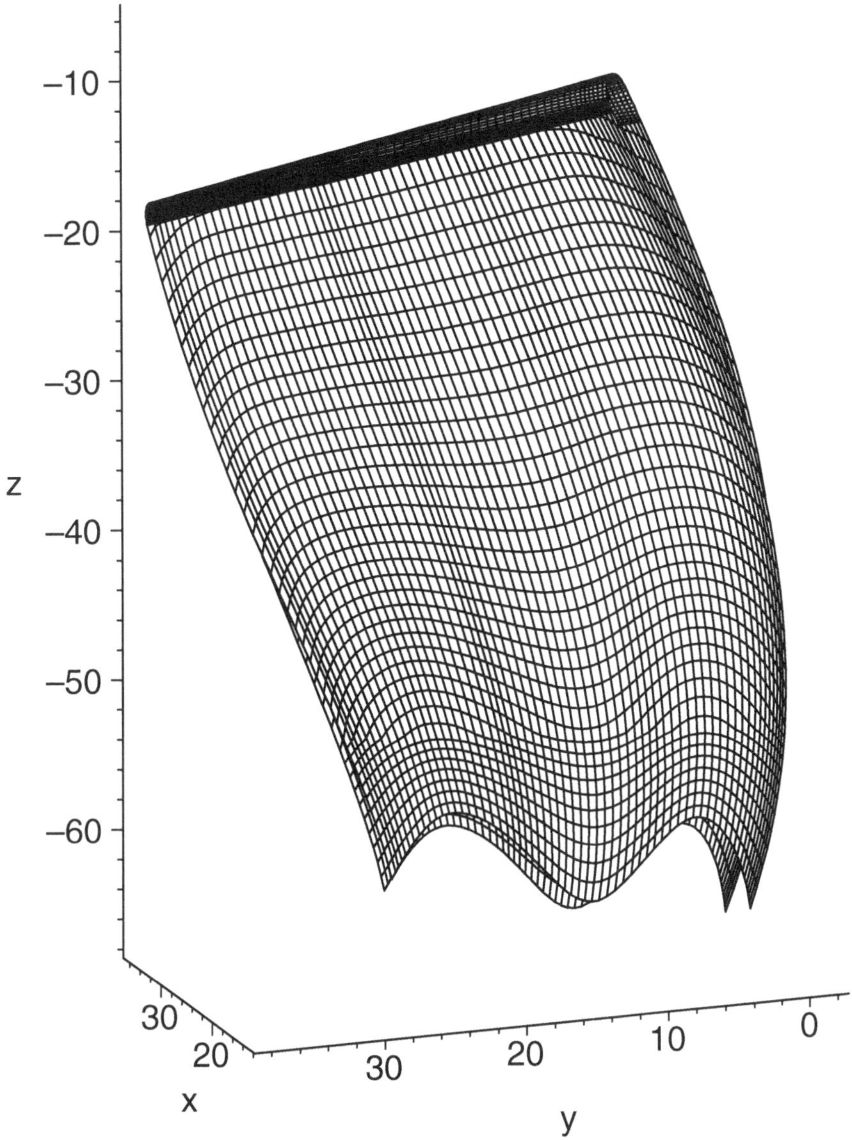

Fig. 5.30. Curvilinear grid adapted to part of a surface of the blade in workpiece coordinate system

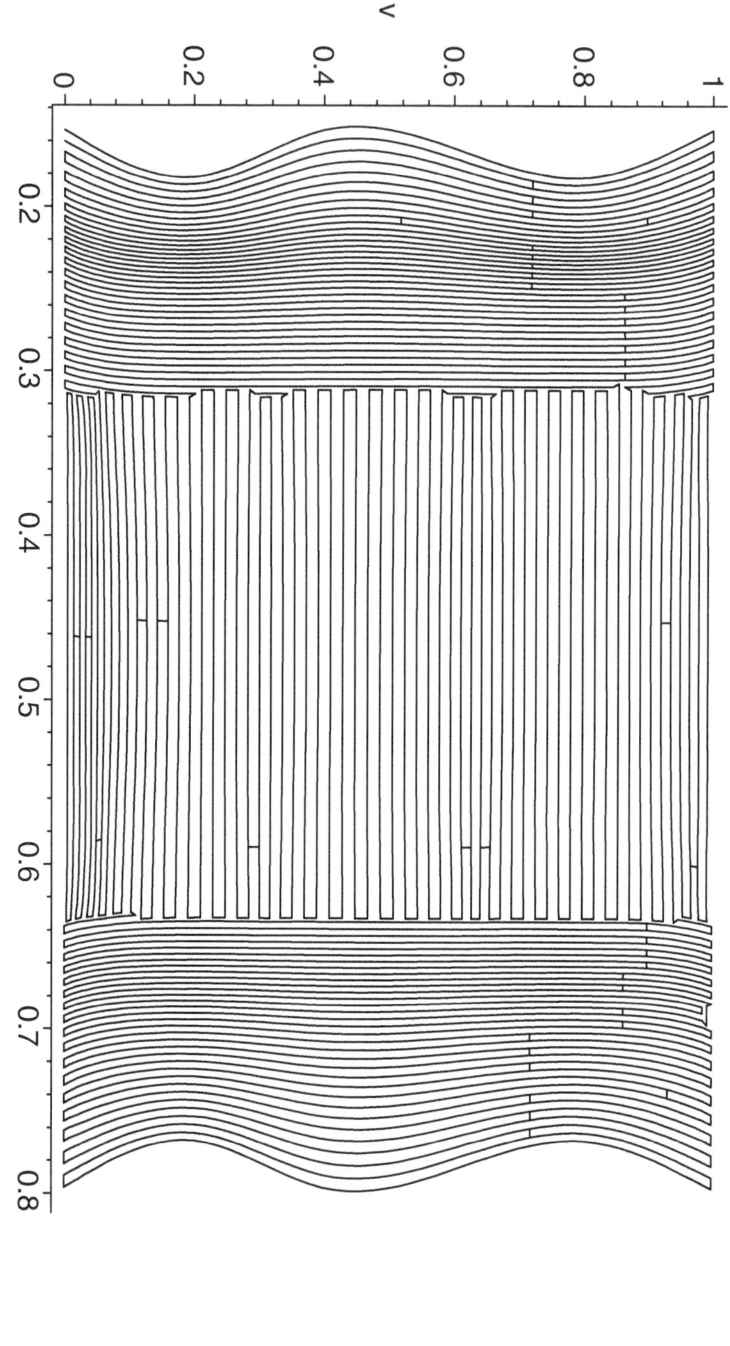

Fig. 5.31. SFC tool path for milling of the broken blade in (u, v) domain

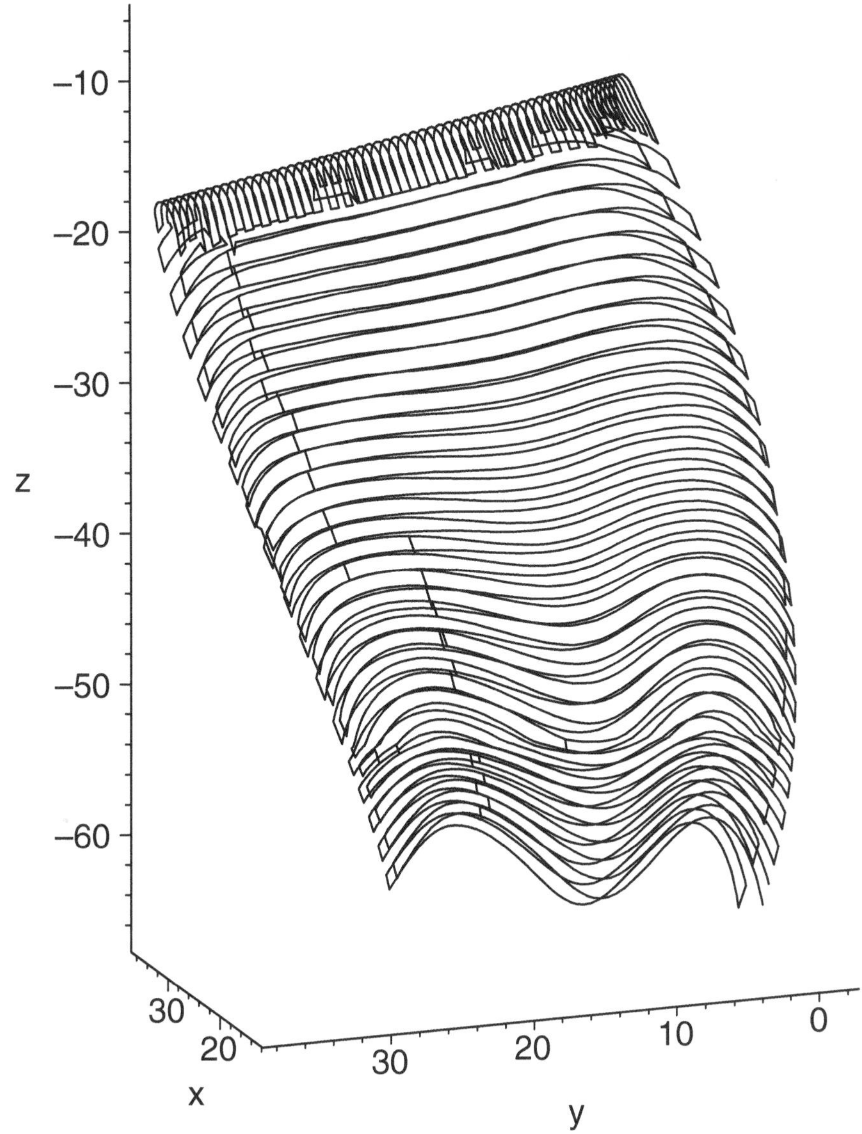

Fig. 5.32. SFC tool path for milling of the broken blade in workpiece coordinate system

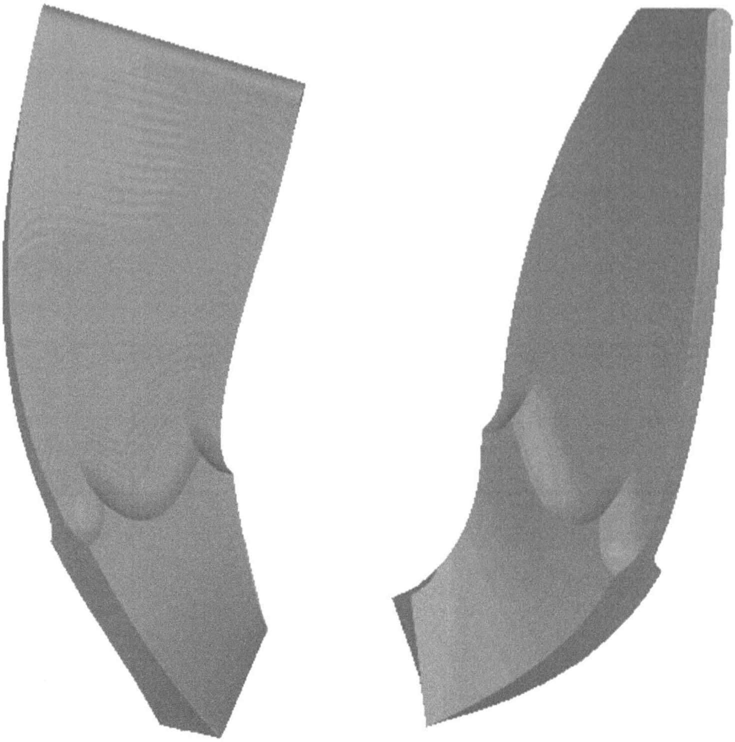

Fig. 5.33. Virtual cutting of the blade in Unigraphics 18

Similarly for node b, the correspondence between local and global node number is given by $n = C_O(N, k)$ and the values of F and its derivatives on u_b and v_b are computed with the use of (5.31) and (5.33). Finally, for node c, the correspondence between local and global node number is given by $n = C_O(N, k+1)$ and the values of F and its derivatives on u_c and v_c are computed with the use of (5.31) and (5.34).

References

[1] Amazigo, J. C. and Rubenfeld, L. A. 1980. *Advanced Calculus and Its Applications to the Engineering and Physical Sciences.* John Wiley & Sons, Inc., New York, USA.

[2] Anderson, D. and Ray, M. 1982. The use of solution adaptive grids in solving partial differential equations. *Applied Mathematics and Computation,* 10-11:317–338.

[3] Anderson, D. A. 1987. Equidistribution schemes, poisson generators, and adaptive grids. *Applied Mathematics and Computation*, 24(3):211–227.

[4] Arina, R. and Tarditi, S. 1994. Orthogonal block structured surface grids. In Baines, M. J. and Morton, K. W., editors, *Numerical Methods for Fluid Dynamics IV*, pages 291–298. Oxford University Press.

[5] Babuska, I. and Rheinboldt, W. 1979. A-posteriori error estimates for the finite element method. *International Journal for Numerical Methods in Engineering*, 12:1597–1615.

[6] Bahvalov, S. 1969. On optimization of methods for solving boundary problems with boundary layers. *Computational Mathematics and Mathematical Physics*, 9(4):841–859.

[7] Baker, T. 1989. Developments and trends in three-dimensional mesh generation. *Applied Numerical Mathematics*, 5(4):275–304.

[8] Berger, M. and Oliger, J. 1984. Adaptive mesh refinement for hyperbolic partial differential equations. *Journal of Computational Physics*, 53(3):484–512.

[9] Bieterman, M. B. and Sandstrom, D. R. 2003. A curvilinear tool-path method for pocket machining. *Journal of Materials Processing Technology*, 125(4):709–715.

[10] Blacker, T. and R.J., M. 1993. Seams and wedges in plastering: A 3D hexahedral mesh generation algorithm. *Engineering with Computers*, 2(9):83–93.

[11] Blacker, T. and Stephenson, M. B. 1991. Paving : a new approach to automated quadrilateral mesh generation. *International Journal for Numerical Methods in Engineering*, 32(4):811–847.

[12] Boender, E. 1994. Reliable Delaunay-based mesh generation and mesh improvement. *Communications in Numerical Methods in Engineering*, 10(10):773–783.

[13] Brackbill, J. U. 1993. An adaptive grid with directional control. *Journal of Computational Physics*, 108(1):38–50.

[14] Brackbill, J. U., B., K. D., and M., R. H. 1988. FLIP: A low-dissipation, particle-in-cell method for fluid flow. *Computational Physics Communications*, 48(1):25–38.

[15] Brackbill, J. U. and Saltzman, J. S. 1982. Adaptive zoning for singular problems in two dimensions. *Journal of Computational Physics*, 46(3):342–368.

[16] Carey, G. 1997. *Computational Grids: Generations, Adaptation & Solution Strategies*. Taylor & Francis, London.

[17] Carey, G. and Dinh, H. 1985. Grading function and mesh redistribution. *SIAM Journal of Numerical Analysis*, 22(3):1028–1040.

[18] Charakhch'yan, A. 1994. Compound difference schemes for time-dependent equations on non-uniform meshes. *Communications in Numerical Methods in Engineering*, 10(2):93–110.

[19] Charakhch'yan, A. A. and Ivanenko, S. A. 1997. A variational form of the Winslow grid generator. *Journal of Computational Physics*, 136(2):385–398.

[20] Dannenhoffer, J. 1988. A comparison of two adaptive grid techniques. In *Numerical Grid Generation in Computational. Fluid Mechanics '88*, pages 319–328.

[21] Dar'in, N. and Majukin, V. . 1988. On one approach to adaptive grid generation in time-dependent problems. *Computational Mathematics and Mathematical Physics*, 28(3):454–460.

[22] De Boor, C. 1974. Good approximation by splines with variable knots II. In Watson, G. A., editor, *Lecture Notes in Mathematics*, volume 363, pages 12–20. Springer-Verlag, Berlin.

[23] Degtyarev, L. and Ivanova, T. 1993. Adaptive grid method in one-dimensional nonstationary convection - diffusion problems. *Differential Equations*, 29(7):1179–1192.

[24] Diaz, A., Kikuchi, N., and Taylor, J. 1983. A method of grid optimization for finite element method. *Computational Methods in Applied Mechanics and Engineering*, 41(1):29–45.

[25] Dvinsky, A. 1991. Adaptive grid generation from harmonic maps on riemanian manifolds. *Journal of Computational Physics*, 95(3):450–476.

[26] Eells, J. and Sampson, J. 1964. Harmonic mappings of riemannian manifolds. *American Journal of Mathematics*, 86(1):109–160.

[27] Eiseman, P. 1987. Adaptive grid generation. *Computational Methods in Applied Mechanics and Engineering*, 64(1-3):321–37664.

[28] Emelyanov, K. 1994. Optimal grid generation and its application in singular problems. *Computational Mathematics and Mathematical Physics*, 34(6):936–943.

[29] Farrashkhalvat, M. and Miles, J. 2003. *Basic Structured Grid Generation: With an Introduction to Unstructured Grid Generation.* Elsevier.

[30] Frey, W. and Field, D. 1991. Mesh relaxation: A new technique for improving triangulation. *International Journal for Numerical Methods in Engineering*, 31(6):1121–1133.

[31] George, P. 1991. *Automatic Mesh Generation: Application to Finite Element Methods.* John Wiley and Sons, New York.

[32] Godunov, S. and Prokopov, G. 1967. On computing conformal mappings and finite-difference grid generation. *Computational Mathematics and Mathematical Physics*, 7(5):1031–1059.

[33] Godunov, S. and Prokopov, G. 1972. The use of moving meshes in gas-dynamics calculations. *Computational Mathematics and Mathematical Physics*, 12(2):182–194.

[34] Golias, N. and Tsiboukis, T. 1994. An approach to refining three-dimensional tetrahedral meshes based on Delauney transformations. *International Journal for Numerical Methods in Engineering*, 37:793–812.

[35] Grebennikov, A. 1976. On the choice of nodes in spline approximation. *Computational Mathematics and Mathematical Physics*, 16(1):219–223.

[36] Ivanenko, S. 1995. Adaptive-harmonic grid generation and its application for numerical solution of the problems with boundary and interior layers. *Computational Mathematics and Mathematical Physics*, 35(10):1203–1220.

[37] Ivanenko, S. A. 1988. Generation of non-degenerate meshes. *Computational Mathematics and Mathematical Physics*, 28(5):141–146.

[38] Ivanenko, S. A. 1993. Adaptive grids and grids on surfaces. *Computational Mathematics and Mathematical Physics*, 33(9):1179–1193.

[39] Ivanenko, S. A. 1999. Harmonic mappings. In *Handbook of Grid Generation*, chapter 8, pages 8(1–43). CRC Press, Inc., Boca Raton, FL, USA.

[40] Jaquotte, O.-P. 1988. A mechanical model for a new grid generation method in computational fluid dynamics. *Computational Mathematics and Applied Mechanics in Engineering*, 66:323–338.

[41] Jeong, J. and Kim, K. 1999. Generation of tool paths for machining free-form pockets with islands using distance maps. *International Journal of Advanced Manufacturing Technology*, 15(5):311–316.

[42] J.F., T., Soni, B., and Weatherill, N. 1999. Handbook of grid generation. *CRC Press.*

[43] Jiang, B. and Carey, G. 1987. Adaptive refinement for least squares finite elements with element-by-element conjugate gradient solution. *International Journal for Numerical Methods in Engineering*, 24(3):569–580.

[44] Kennon, S. and Anderson, D. 1988. Unstructured grid adaption for non-convex domains. In *Proceedings of the Conference on Numerical Grid Generation in Computational Fluid Mechanics*, pages 599–609.

[45] Kennon, S. and Dulikravich, G. 1986. Generation of computational grids using optimization. *AIAA Journal*, 24(7):1069–1073.

[46] Liseikin, V. 2006. *A Computational Differential Geometry Approach to Grid Generation.* Springer,, Berlin, New York.

[47] Liseikin, V. D. 1999. *Grid Generation Methods.* Springer-Verlag, Berlin, Germany.

[48] Lo, C. C. 1999. Efficient cutter-path planning for five-axis surface machining with a flat-end cutter. *Computer-Aided Design*, 31(9):557–566.

[49] Makhanov, S. S. 1999. An application of variational grid generation techniques to the tool-path optimization of industrial milling robots. *Journal of Materials Processing Technology*, 39(9):1524–1535.

[50] Makhanov, S. S., Batanov, D., Bohez, E., Sonthipaumpoon, K., Anotaipaiboon, W., and Tabucanon, M. 2002. On the tool-path optimization of a milling robot. *Computers & Industrial Engineering*, 43(3):455–472.

[51] Makhanov, S. S. and Ivanenko, S. A. 2003. Grid generation as applied to optimize cutting operations of the five-axis milling machine. *Applied Numerical Mathematics*, 46(3-4):331–351.

[52] Muller, K. 1981. Moving finite elements II. *SIAM Journal of Numerical Analysis*, 18(6):1033–1057.

[53] Muller, K. and Muller, R. 1981. Moving finite elements I. *SIAM Journal of Numerical Analysis*, 18(6):1019–1032.

[54] Nakahashi, K. and Deiwert, G. S. 1986. Three-dimensional adaptive grid method. *AIAA Journal*, 24(6):948–045.

[55] Oden, J., Demkowicz, L., Strouboulis, T., and Devlow, P. 1986. Adaptive methods for problems in solid and fluid mechanics. In Babuska, I., editor, *Accuracy Estimates and Adaptive Refinements in Finite Element Computations*, pages 249 – 280. John Willey.

[56] Pardhanani, A. and Carey, G. 1988. Optimization of computational grids. *Numerical Methods for Partial Differential Equations*, 4(2):95–117.

[57] P.D., T. 1982. Composite three dimensional grids generated by elliptic systems. *AIAA Journal*, 20(9):1195–1202.

[58] Pereira, V. and Sewell, E. 1975. Mesh selection for discrete solutions of boundary value problems in ordinary differential equations. *Numerical Mathematics*, 23(3):261–268.

[59] Prokopov, G. 1970. Construction of orthogonal difference grids from conformal mappings. In *Proceedings N45, Keldysh Institute of Applied Mathematics*, Moscow (in Russian).

[60] Rank, E. and Babushka, I. 1987. An expert system for the optimal mesh design in hp-version of the finite element method. *International Journal for Numerical Methods in Engineering*, 24(11):2087–2106.

[61] Roache, P. and Steinberg, S. 1985. A new approach to grid generation using a variational formulation. In *Proceeding of AIAA 7-th Computational Fluid Dynamics Conference*, pages 360–370.

[62] Russel, R.D.and Christiansen, J. 1978. Adaptive mesh selection strategies for solving boundary value problems. *SIAM Journal of Numerical Analysis*, 15(1):59–80.

[63] Ryskin, G. and L.G., L. 1983. Orthogonal mapping. *Journal of Computational Physics*, 50(1):71–100.

[64] Saltzman, J. 1986. Variational methods for generating meshes on surfaces. *Journal of Computational Physics*, 63(1):1–19.

[65] Samareh-Abolhassani, J. and Stewart, J. 1994. Surface grid generation in a parameter space. *Journal of Computational Physics*, 113(1):112–121.

[66] Serezhnikova, T., Sidorov, A., and Ushakova, O. 1989. On one method of construction of optimal curvilinear grids and its applications. *Soviet Journal Numerical Analysis and Mathematical Modelling*, 4(2):137–155.

[67] Sidorov, A. 1966. On one algorithm for computing optimal difference grids. In *Proceedings of the Steklov Institute of Mathematics*, pages 147–151.

[68] Soni, B. 1993. Elliptic grid generation system: Control functions revisited - I. *Applied Mathematics and Computation*, 59:151–163.

[69] Spitaleri, R. and Micacchi, V. 1998. A multiblock multigrid generation method for complex simulations. *Mathematics and Computers in Simulation*, 46(1):1–12.

[70] Steger, J. and Chaussee, D. 1980. Generation of body-fitted coordinates using hyperbolic partial differential equations. *SIAM Journal on Scientific and Statistical Computing*, 1(4):431–437.

[71] Steinberg, S. 1994. *Fundamentals of Grid Generation.* CRC Press.

[72] Thompson, J. 1984. Grid generation techniques in computational fluid dynamics. *AIAA Journal*, 22(11):1505–1523.

[73] Thompson, J., Warsi, Z., and .Mastin, C. 1985. *Numerical grid generation: Foundations and Applications.* North-Holland.

[74] Thompson, J. F., editor 1982. *Numerical Grid Generation.* North-Holland.

[75] Tikhonov, A. and Gorbunov, A. 1969. Error estimates for the Runge-Kutta methods and choice of optimal grids. *Computational Mathematics and Mathematical Physics*, 4(2):232–242.

[76] Tu, Y. and Thompson, J. 1991. Three-dimensional solution - adaptive grid generation on composite configurations. *AIAA Journal*, 29(12):2025–2026.

[77] Tucker, A. B., editor 1997. *The Computer Science and Engineering Handbook.* CRC Press.

[78] Vabistchevitch, P. 1989. Adaptive block grids in problems of mathematical. *Computational Mathematics and Mathematical Physics*, 29(6):902–914.

[79] White, A. J. 1979. On selection of equidistributing meshes for two-point boundary-value problems. *SIAM Journal of Numerical Analysis*, 16(3):472–502.

[80] White, A. J. 1982. On the numerical solution of initial/boundary value problems in one space dimension. *SIAM Journal of Numerical Analysis*, 19(4):683–697.

[81] Winslow, A. M. 1966. Numerical solution of the quasilinear Poisson equation in a nonuniform triangle mesh. *Journal of Computational Physics*, 1(2):149–172.

6

Optimization of Rotations

6.1 Introduction

Given the general optimization context given in the preceding chapters, we will now analyze a particular but important problem of sequencing the machine rotations. There exists a variety of research focused on the *orientation of the cutting tool.* The survey of which is given in Chap. 1. However, the accuracy is also affected by *the way the orientations are being achieved.* In other words, the kinematics error depends not only on the characteristics of the surface versus the tool orientation but on the previous rotations as well. The "history" of rotations becomes in particular important in the vicinity of the stationary points of the desired surface where the tool could make abrupt changes in the orientation. However, such analysis is not provided by commercial CAD/CAM systems such as Unigraphics, EdgeCam, Vericut, etc. Besides, only a few recent research papers deal with the subject.

Recall that a stationary point on a surface is either a maximum or a minimum point or a saddle point. Mathematically, it means that if (u_s, v_s) is the stationary point then

$$\left.\frac{\partial S}{\partial u}\right|_{(u_s,v_s)} = \left.\frac{\partial S}{\partial v}\right|_{(u_s,v_s)} = 0. \tag{6.1}$$

Jung et al. [4] analyze the sequence of rotations to minimize the number of the phase reverse steps at discontinuities of the first derivative of the surface (corners, etc.). A method of avoiding singularities has been presented by Affouard et al. [1]. An algorithm based on global optimization with regard to feasible rotations of the machine was proposed in [11] and developed in [9].

The optimization is based on the idea that there exist several combinations of rotation angles to establish the required orientation. Therefore, one can formulate a problem of finding a set of the rotation angles sequenced in such a way that a certain cost function (for instance the kinematics error) is reduced. The problem can be solved by the shortest path optimization with regard to the feasible rotations found from the kinematics equations. The cost function can be represented in terms of either the kinematics error or the angle variation. Such optimization increases the accuracy of machining and is the most appropriate in the case of a rough cut.

The shortest path procedure applies to either the entire set of trajectories or to only the most inappropriate overcuts inside the workpiece. In the latest case the algorithm generates an interesting family of solutions characterized by smaller overcuts obtained at the expense of increased undercuts.

Note that a fine cut of a smooth surface employing small spatial and angular steps may not demonstrate the detrimental effects near the stationary points. However, a rough cut characterized by large gradients could produce considerable errors.

The large gradients lead to sharp angular jumps and it is because of the sharp angular jumps that the machine produces the loop-like trajectories of the tool. Moving along such trajectories could destroy the workpiece and even lead to collisions with the machine parts.

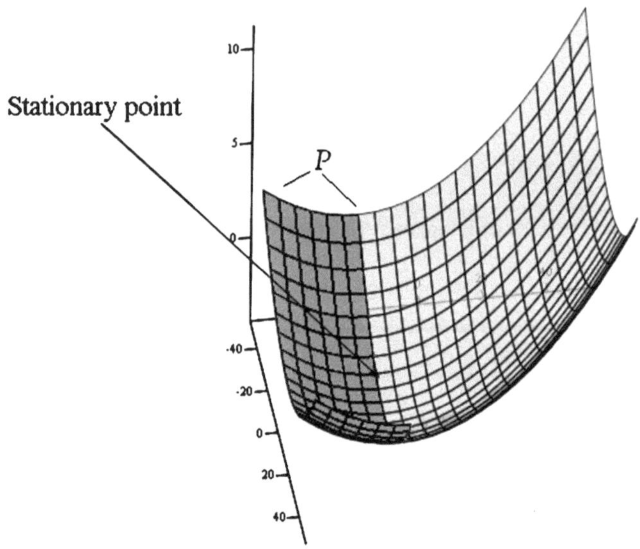

Fig. 6.1. An experimental part surface S_1 (*P is the machined area*)

Our introductory example presents a cut performed for a part surface S_1 (Fig. 6.1) on HERMLE UWF902H. Figure 6.2 shows a large loop in the case of machining a curve belonging to the surface located close to a stationary point. Clearly, linearization of the tool path is not always applicable. It could be even unsafe to assume the linear trajectories, since moving the tool along the actual loop-like trajectories could destroy the workpiece and lead to collisions with the machine parts.

Figure 6.3 shows that changing the sequence of the rotations leads to a much lesser loop and consequently to a significant reduction of the kinematics error. Our second example demonstrates a similar trajectory when machining a single curve belonging to surface S_2 (Fig. 6.4) on MAHO 600E.

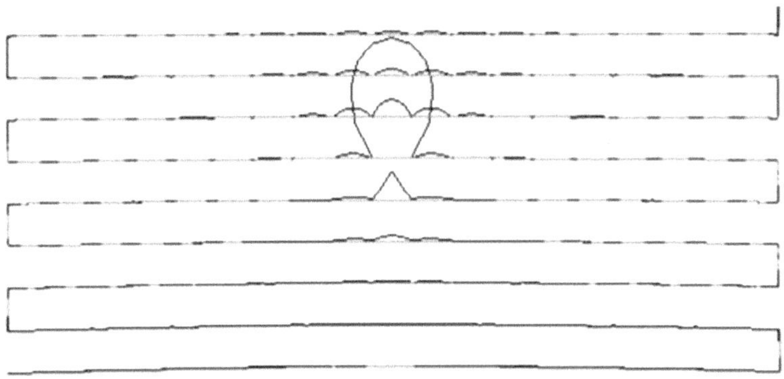

(a)

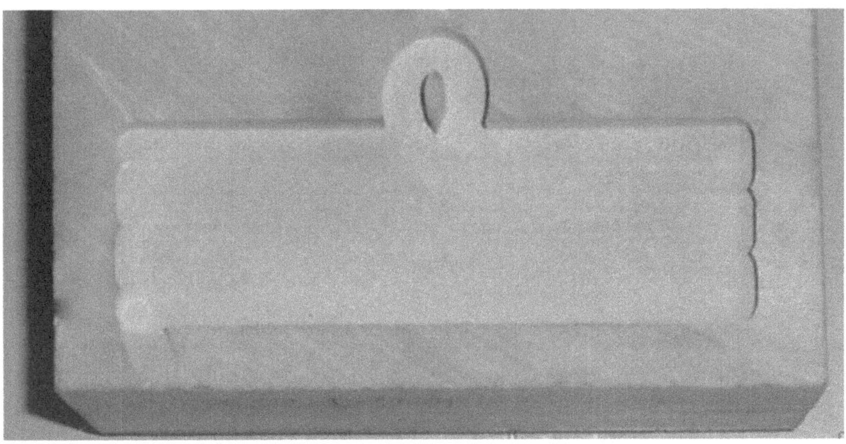

(b)

Fig. 6.2. (**a**) conventional tool path simulated by the virtual machine, (**b**) surface machined by HERMLE UWF902H

Figure 6.5 shows that as opposed to a linearized version of the tool path, the real machining produces a loop-like trajectory induced by the large angular steps. Such trajectories could be repaired (see Fig. 6.6) by adjusting the rotation angles in such a way that the kinematics error is minimized.

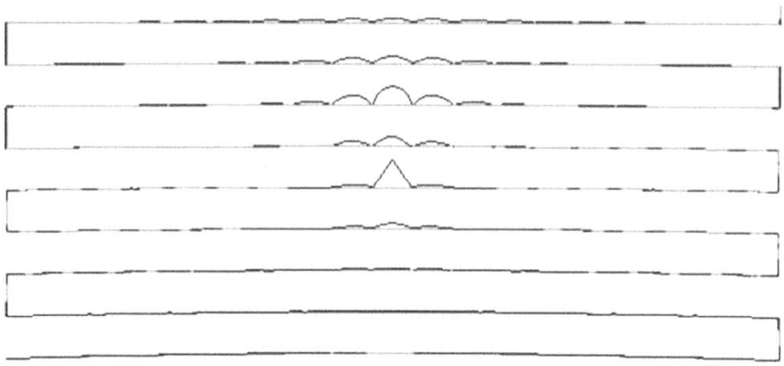

(a)

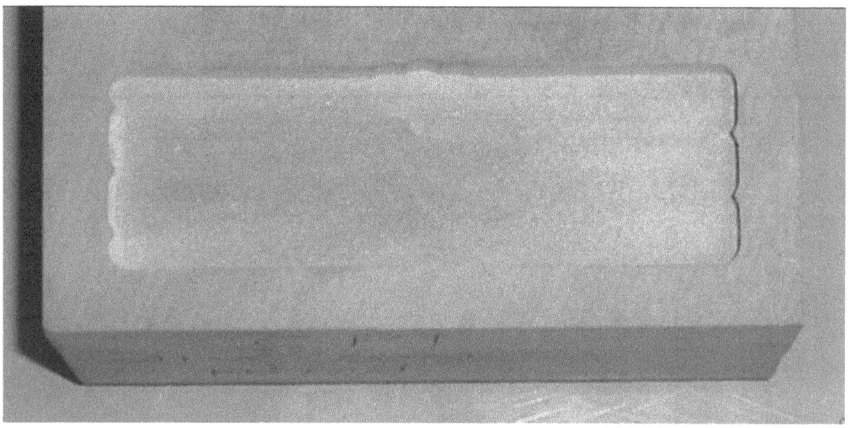

(b)

Fig. 6.3. (**a**) optimized tool path simulated by the virtual milling machine, (**b**) optimized surface machined by HERMLE UWF902H

In this chapter, we will also introduce a point inserting algorithm which can be interpreted as an angular interpolation scheme in the areas of large kinematics errors. The additional positions are created by finding numerically a grid of points uniformly distributed in the angular space.

The procedures are in particular beneficial for high speed milling characterized by small spatial steps. In this case a further decrease of the step size leads to a substantial increase in the machining time due to delays in the servo-update rate.

Finally, there is always a limit of the angular speed of specific machine parts. As a result, a shorter tool path with many turns may require more time than a longer tool path with fewer turns [2, 3]. If the maximum angular speed is exceeded, the controller detects this event and reduces the angular speed

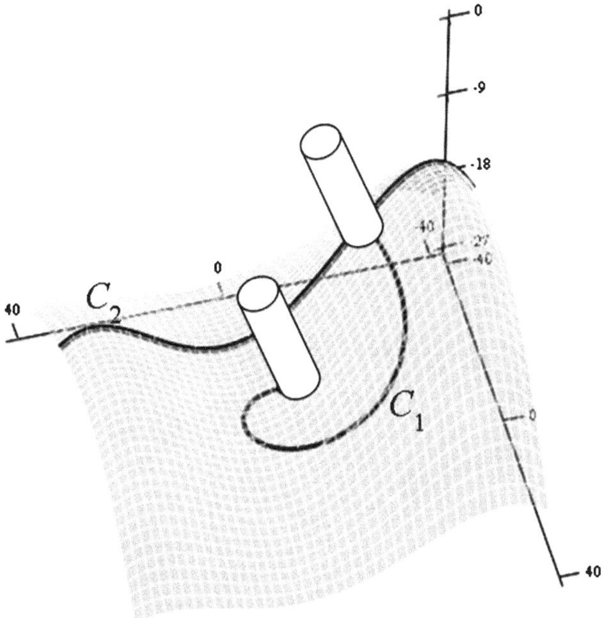

Fig. 6.4. Nonlinearity of the tool path due to rotations in the workpiece coordinates (C_1), the experimental cut (C_2)

increasing the machining time. The angle correction algorithms minimize the total angle variation, thus, reducing the probability of such an event.

6.2 Kinematics Error and Angle Variation

Let $W^D_{p,p+1}(t) \in S(u,v)$ be a curve between W_p and W_{p+1} extracted from the surface in such a way that it represents a desired tool trajectory between Π_p and Π_{p+1} (see also Sect. 3.4). We define the *total error* as a sum of deviations of $W^D_{p,p+1}(t)$ from the actual trajectories $W_{p,p+1}(t)$. The total error is given by

$$\begin{aligned} \epsilon &= \sum_p \left[\int_0^1 \left| W^D_{p,p+1}(t) - W_{p,p+1}(t) \right|^2 dt \right]^{\frac{1}{2}}, \\ &= \sum_p \left[\int_0^1 \left[\left(x^D_{p,p+1}(t) - x_{p,p+1}(t) \right)^2 + \left(y^D_{p,p+1}(t) - y_{p,p+1}(t) \right)^2 \right. \right. \\ &\quad \left. \left. + \left(z^D_{p,p+1}(t) - z_{p,p+1}(t) \right)^2 \right] dt \right]^{\frac{1}{2}}. \end{aligned} \tag{6.2}$$

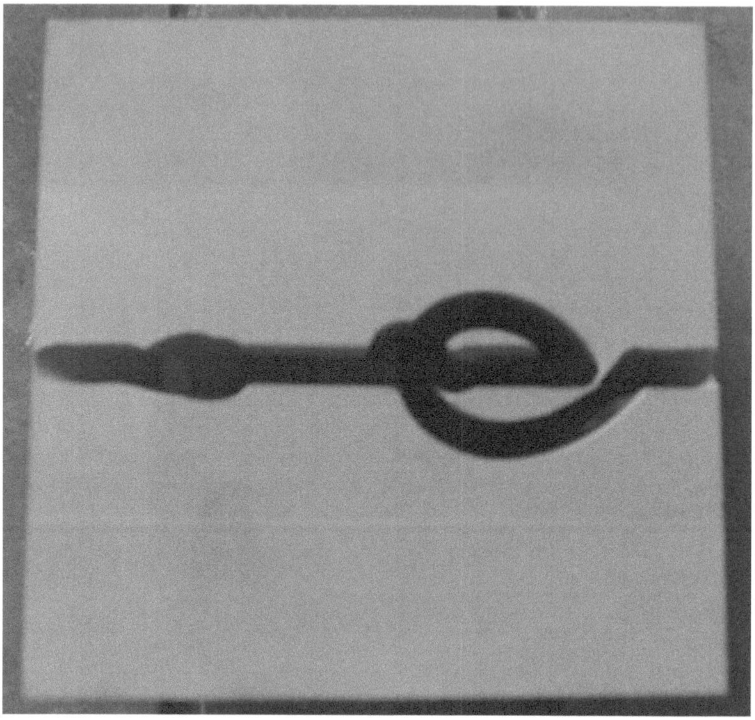

(a)

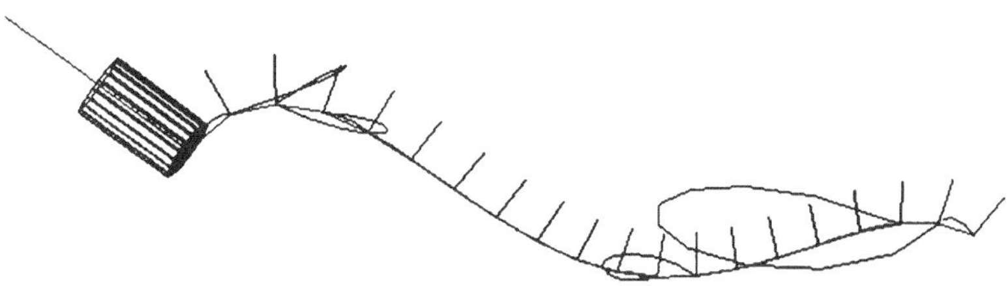

(b)

Fig. 6.5. (**a**) a loop-like trajectory induced by large gradients of the rotation angles damages the workpiece, (**b**) the trajectory simulated by the virtual milling machine

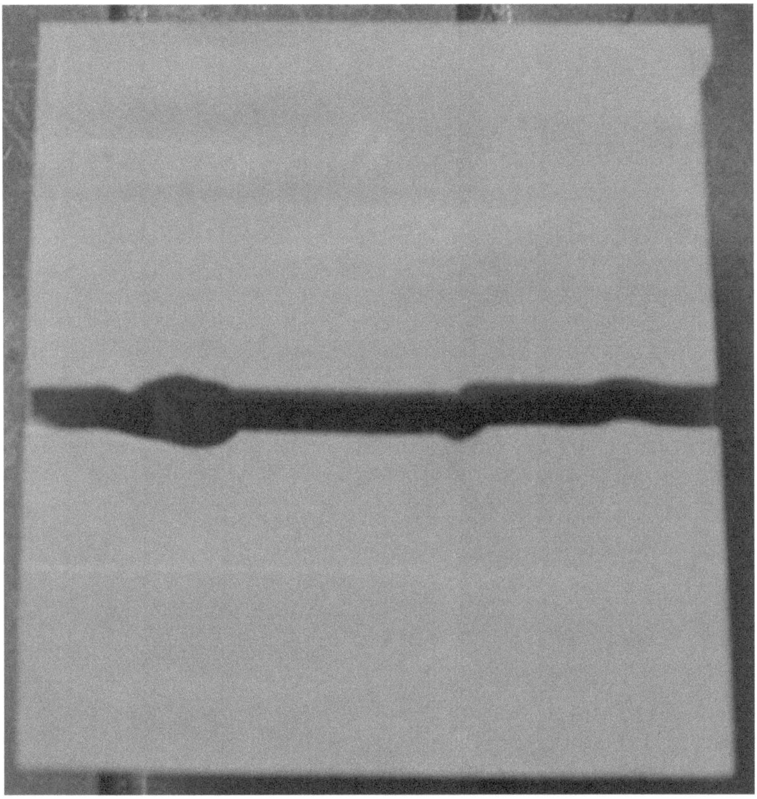

(a)

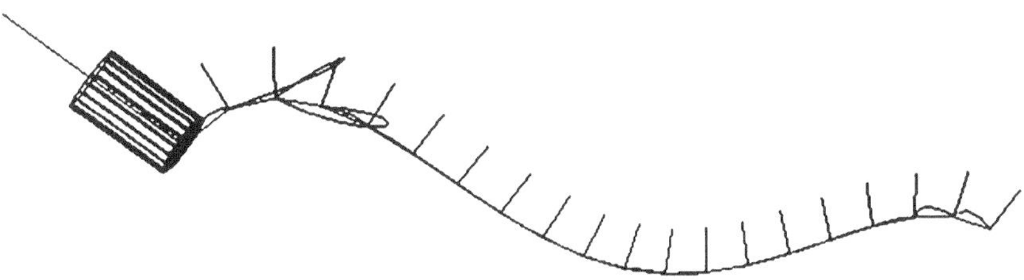

(b)

Fig. 6.6. (**a**) the "repaired" trajectory, (**b**) the repaired trajectory simulated by the virtual milling machine

The total error (6.2) is approximated as follows:

$$\epsilon = \sum_p \frac{1}{L_p} \left[\sum_{l_p=1}^{L_p} \left[\left(x^D_{p,p+1,l_p} - x_{p,p+1,l_p} \right)^2 + \left(y^D_{p,p+1,l_p} - y_{p,p+1,l_p} \right)^2 + \left(z^D_{p,p+1,l_p} - z_{p,p+1,l_p} \right)^2 \right] \right]^{\frac{1}{2}}, \tag{6.3}$$

where L_p is a number of the sampling points between W_p and W_{p+1}.

Remark 6.1. The definitions above can be simplified by replacing the desired trajectories $W^D_{p,p+1}$ by linear trajectories given by $W^L_{p,p+1}(t) = tW_{p+1} + (1-t)W_p$, although care should be taken when using this option. As opposed to the machine coordinates M, the trajectories in the workpiece coordinates W are not linear. However, we may use the linear trajectories as a reference, noting that $\epsilon \equiv \|W^D - W\| \leq \|W^D - W^L\| + \|W^L - W\|$, where L is a piecewise linear approximation of S. Hence, when the points are close enough, the error is approximately $\epsilon \approx \epsilon_L = \|W^L - W\|$. Note that if the orientation of the tool is fixed through the entire cut, then $\epsilon_L = 0$. In other words, the three-axis mode leads to $\epsilon_L = 0$ since all the trajectories become linear. Therefore, minimization with regard to ϵ_L must be subjected to constraints specifying the orientations of the tool. Finally, the linearization is the simplest option which can be used even when the actual surface is not known, for instance, for a given G-code. Actually, the desired trajectory W^D can be extracted from the surface by a variety of ways, for example, using interpolation in the parametric space, the geodesic curves, etc. However, the results presented in this chapter are valid irrespectively of the method to obtain W^D. Finally, the error induced by linear approximation depends on the maximum distance h between the CC points as $k\|\frac{d^2W^D}{dt^2}\|_C h^2$, where k is a constant (see, for instance, [10]).

Let us now introduce the total angle variation. Consider two positions W_p and W_{p+1}. The corresponding kinematics error between the two positions is then defined by

$$\epsilon_{p,p+1} \equiv \left[\int_0^1 \left[W^D_{p,p+1}(t) - W_{p,p+1}(t) \right]^2 dt \right]^{\frac{1}{2}}. \tag{6.4}$$

Clearly, $\epsilon_{p,p+1} \equiv \epsilon(\Delta a_{p,p+1}, \Delta b_{p,p+1}, \Delta l_{p,p+1})$ where $\Delta l_{p,p+1} = \|W^D_{p,p+1} - W_{p,p+1}\|_E$ is the spatial step in the workpiece coordinate system. Furthermore, $\Delta a_{p,p+1} \equiv a_{p,p+1}(\Delta l_{p,p+1}) = a_{p+1} - a_p$ and $\Delta b_{p,p+1} \equiv b_{p,p+1}(\Delta l_{p,p+1}) = b_{p+1} - b_p$ are the angular steps. Machining experiments show that the angular steps are often more important than the spatial step. In particular, during a rough cut, decreasing the angular steps leads to a larger decrease in the error than a decrease in the spatial step. The angle variation between positions W_p and W_{p+1} is given by:

$$c_{p,p+1} = \int_0^1 \sqrt{\left(\frac{\partial a_{p,p+1}}{\partial t}\right)^2 + \left(\frac{\partial b_{p,p+1}}{\partial t}\right)^2} dt. \tag{6.5}$$

Assuming that $a_{p,p+1}$ and $b_{p,p+1}$ are changing linearly, yields $c_{p,p+1} = \sqrt{(\Delta a_{p,p+1})^2 + (\Delta b_{p,p+1})^2}$. The total angle variation across the entire tool path is then simply given by

$$c = \sum_p c_{p,p+1} = \sum_p \sqrt{(\Delta a_{p,p+1})^2 + (\Delta b_{p,p+1})^2}. \tag{6.6}$$

Note that there often exists a neighborhood of the CL point where one of the angles changes faster than another one. In this case $\epsilon_{p,p+1} \approx O\left((\Delta a_{p,p+1})^{m_1}\right)$ or $\epsilon_{p,p+1} \approx O\left((\Delta b_{p,p+1})^{m_2}\right)$, where m_1 and m_2 depend on the surface. In this case, minimization of the total angle variation produces excellent results.

Finally, the case of a rough cut often requires to differentiate between the overcut and undercut error. When too much material is removed from the part, it is said to be gouged or overcut. The undercut is a situation when not enough material is removed. For simplicity, we define the undercut and overcut error as follows

$$\epsilon_U = \sum_p \int_0^1 \begin{cases} (W^D_{p,p+1} - W_{p,p+1})^2 & \text{if } z^D_{p,p+1}(t) \leq z_{p,p+1}(t) \\ 0 & \text{otherwise} \end{cases} dt\ , \tag{6.7}$$

$$\epsilon_O = \sum_p \int_0^1 \begin{cases} (W^D_{p,p+1} - W_{p,p+1})^2 & \text{if } z^D_{p,p+1}(t) > z_{p,p+1}(t) \\ 0 & \text{otherwise} \end{cases} dt\ . \tag{6.8}$$

The corresponding undercut and overcut angle variation are then defined by

$$c_U = \sum_p c_{U,p,p+1}, \tag{6.9}$$

$$c_O = \sum_p c_{O,p,p+1}, \tag{6.10}$$

where

$$c_{U,p,p+1} = \sqrt{(\Delta a_{p,p+1})^2 + \Delta b_{p,p+1})^2 \Delta^U_{p,p+1}},$$

$$c_{O,p,p+1} = \sqrt{(\Delta a_{p,p+1})^2 + \Delta b_{p,p+1})^2 \Delta^O_{p,p+1}},$$

$$\Delta^U_{p,p+1} = \int_0^1 \begin{cases} 1 \text{ if } z^D_{p,p+1}(t) \leq z_{p,p+1}(t) \\ 0 \text{ otherwise} \end{cases} dt\ ,$$

$$\Delta^{O}_{p,p+1} = \int_0^1 \begin{cases} 1 \text{ if } z^D_{p,p+1}(t) > z_{p,p+1}(t) \\ 0 \text{ otherwise} \end{cases} dt ,$$

Clearly, $\epsilon = \epsilon_O + \epsilon_U$ and $c = c_O + c_U$. Since overcut is not repairable, reducing the overcut error is often more important than reducing the undercut or even the total error.

6.3 Optimization Problem

Let us formulate the following minimization problem:

$$\min_{\Lambda}(w_U \epsilon_U + w_O \epsilon_O), \tag{6.11}$$

where Λ is a set of possible rotations and w_U, w_O the weighting coefficients representing the importance of the undercut and overcut, respectively. For instance, if $w_U = w_O = 1$, then the total kinematics error will be minimized. If $w_U = 0$, $w_O = 1$, only kinematics errors along the overcut will be minimized. If $w_U = 0.1$, $w_O = 1.0$, then the kinematics error is minimized with regard to both the undercut and overcut, but the importance of the overcut is 10 times larger than that of the undercut. The weights may also depend on the kinematics error itself. For example, one may establish the following rule "if $\epsilon < \epsilon'$ then $w_O = 1$, $w_U = 0$ else $w_O = w_U = 1$", where ϵ' is a prescribed threshold. In other words, if the total kinematics error is not too large, we minimize only the overcut, however if the kinematics error is more than a prescribed ϵ' then the minimization applies to the total kinematics error.

How to evaluate set Λ? Without loss of generality, consider the configuration depicted in Fig. 2.4. Recall that the machine movements prescribed by a sequence of the machine commands are executed simultaneously allowing for five degrees of freedom of the tool. The goal of the simultaneous movements is transporting the tool tip to location $W = (x_w, y_w, z_w)$ in the workpiece coordinate system or $M = (x_m, y_m, z_m)$ in the machine coordinate systems and establishing orientation $\mathfrak{R} = (a, b)$. In accordance with the reference coordinate system in Fig. 2.4(b), the angle a is between the X_1-axis and the projection of the orientation vector onto the X_1-Y_1 plane and the positive direction of the X_1-axis. The angle b is between the projection of the orientation vector onto the X_1-Z_1 plane and the positive direction of the X_1-axis.

The above makes it possible to represent $\mathfrak{R}$ in terms of the components of the orientation vector $I = (I_x, I_y, I_k)$. First of all, note that the tool orientation vector in O_4 is $(0, 0, 1)$. In other words, the rotations must be performed in such a way that the tool orientation vector becomes collinear with the Z_4-axis. Observe that the resulting coordinate transformation depends on the zero position $\mathfrak{R} = (0, 0)$ which can assigned by a special machine command. It is not hard to see clearly that at this configuration only $b = 0$ is important. Therefore, without loss of generality, we assign $\mathfrak{R} = (0, 0)$ to the rightmost position of the tilt table shown in Fig. 2.4(a).

From (2.4), the equation relating the two rotation angles $\mathfrak{R} = (a, b)$ and the tool orientation vector I is given by

$$\begin{aligned} I_x &= \cos(a)\cos(b), \\ I_y &= \sin(a)\cos(b), \\ I_z &= -\sin(b). \end{aligned} \tag{6.12}$$

Consider a solution of the above system given by

$$\begin{aligned} a_{\text{base}} &= \begin{cases} \tan^{-1}\left(\dfrac{I_y}{I_x}\right) & \text{if } I_x > 0 \text{ and } I_y \geq 0, \\ \tan^{-1}\left(\dfrac{I_y}{I_x}\right) + \pi & \text{if } I_x < 0, \\ \tan^{-1}\left(\dfrac{I_y}{I_x}\right) + 2\pi & \text{otherwise,} \end{cases} \\ b_{\text{base}} &= -\sin I_z, \end{aligned} \tag{6.13}$$

Given a_{base} and b_{base}, it is not hard to construct a full set of four solutions given by (see the four solutions in Fig. 6.7)

$$\begin{aligned} \Lambda = \big\{ &(a_{\text{base}}, b_{\text{base}}), (a_{\text{base}} - 2\pi, b_{\text{base}}), \\ &(a_{\text{base}} - \pi, -b_{\text{base}} - \pi), (a_{\text{base}} + \pi, -b_{\text{base}} - \pi) \big\}. \end{aligned} \tag{6.14}$$

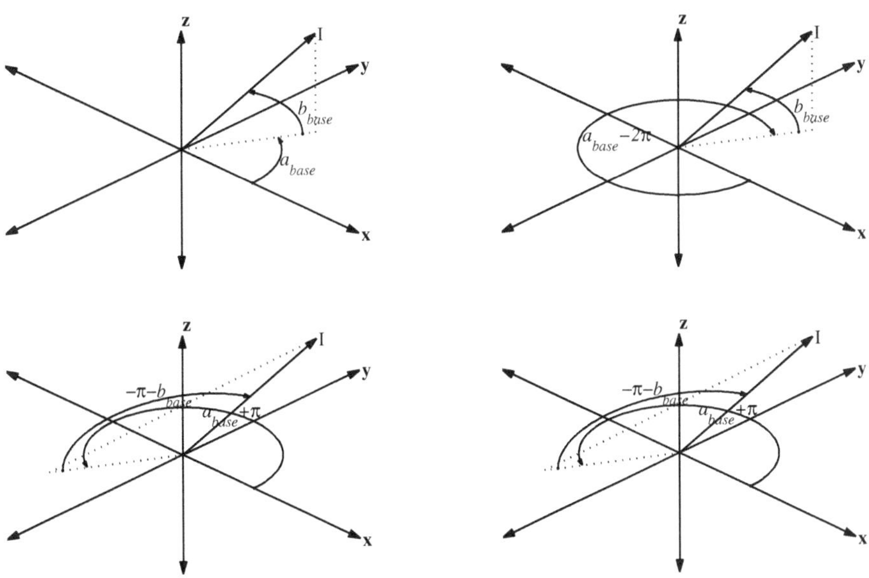

Fig. 6.7. The set of feasible rotations

Similar solutions can be established for other types of the machine kinematics such as, the 1-1 machine and the 0-2 machine (see Chap. 2).

Note that the number of solutions depends on the number of the rotational degrees of freedom d as $O(2^d)$. For example, for a six-axis machine we will have eight solutions, however, some solutions might not be achievable due to the machine limits.

Furthermore, it is clear that the problem can be solved by shortest path optimization. The corresponding graph is constructed in such a way that each position Π_p is characterized by 4 graph nodes Λ_p, where the edge between the nodes represents the kinematics error or the angle variation (see Fig. 6.8). Therefore, such minimization could be performed by a conventional so-called greedy discrete algorithm.

The above shortest path scheme is visualized in Fig. 6.9 and Fig. 6.10, where l denotes a coordinate along the tool path. The 4 options are represented by 4 trajectories in the angular space (a, b). It is plain that the trajectories are close when b is near $-\pi/2$. In this case it is possible to change the trajectory by considering the shortest path producing minimal error. Figure 6.10 shows that the optimization may only be required near the stationary points, namely, when b is close to $-\pi/2$ and a is "jumping" from 0 to π or from $\pi/2$ to $3\pi/2$, etc. Consequently, the computational load could be substantially reduced by introducing an "optimization windows" in the neighborhood of the stationary points. This idea discussed in details in [8].

6.4 Optimization Problems: Examples and Practical Machining

Example 6.1. **A trajectory passing through a stationary point - across or around the hill**

Consider two successive points $\Pi_1 = (W_1, \mathfrak{R}_1)$ and $\Pi_2 = (W_2, \mathfrak{R}_2)$. Let the average space step $s = |W_1 - W_2|$ be 1 mm. Consider a fixed $\mathfrak{R}_1 \equiv (a_1, b_1) = (0, -85°)$ and a varying $\mathfrak{R}_2 \in [120, 180] \times [-60, -80]$. $\mathfrak{R}_1$ and $\mathfrak{R}_2$ represent a possible combination of the rotation angles in the neighborhood of a stationary point (see Fig. 6.11).

Consider a linear trajectory T and the actual trajectories T_{across}, T_{around} corresponding to $\mathfrak{R}_2 \equiv (a_2, b_2) = (-10°, -95°)$ and $\mathfrak{R}_{2,\text{base}} \equiv (a_{2,\text{base}}, b_{2,\text{base}}) = (170, -85)$, respectively. Note that $a_2 = a_{2,\text{base}} - 180°$, $b_2 = -b_{2,\text{base}} - 180°$. Let the distance between the midpoint of the linear trajectory and the center of the a-rotation (A-axis) r_a be 10 mm. Let us compute the corresponding trajectory and the kinematics error. We have $\epsilon_{\text{across}} = 1.49$ mm, $\epsilon_{\text{around}} = 8.89$ mm. In words, moving "across the hill" by T_{across} is better than T_{around} - "around the hill".

However, it is not always the case. A small r_a may entail another choice. For instance, $r_a = 0.5$ mm produces $\epsilon_{\text{around}} = 0.71$ mm, $\epsilon_{\text{across}} = 1.59$ mm.

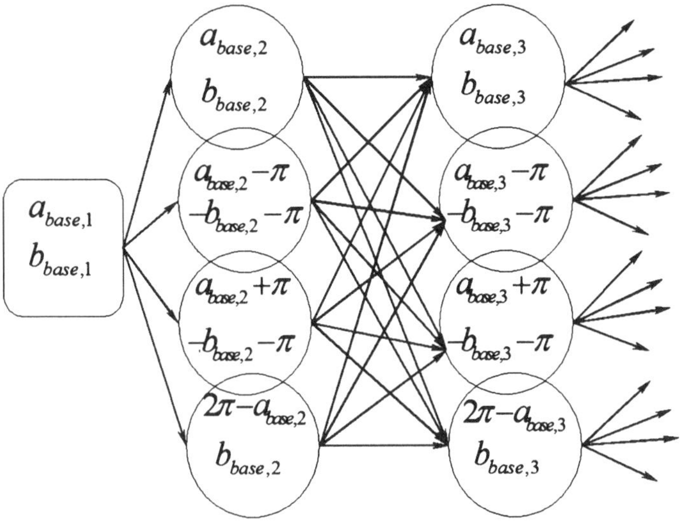

Fig. 6.8. A graph corresponding to the set of the feasible rotations

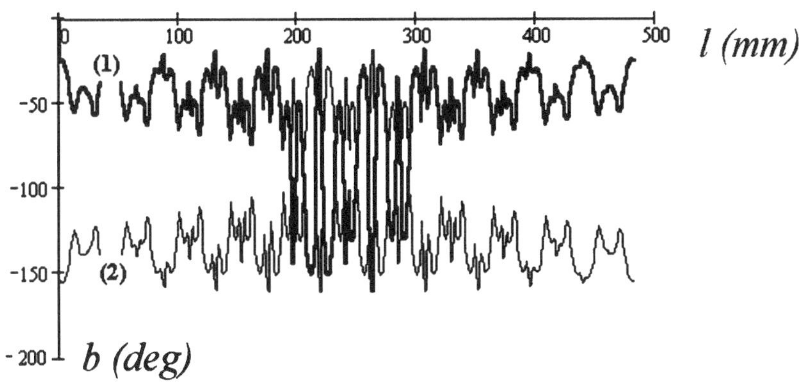

Fig. 6.9. Trajectories corresponding to (1) b_{base} and (2) $\pi - b_{\text{base}}$ and the optimized trajectory composed from (1) and (2) (*bold line*), l is the coordinate along the tool path

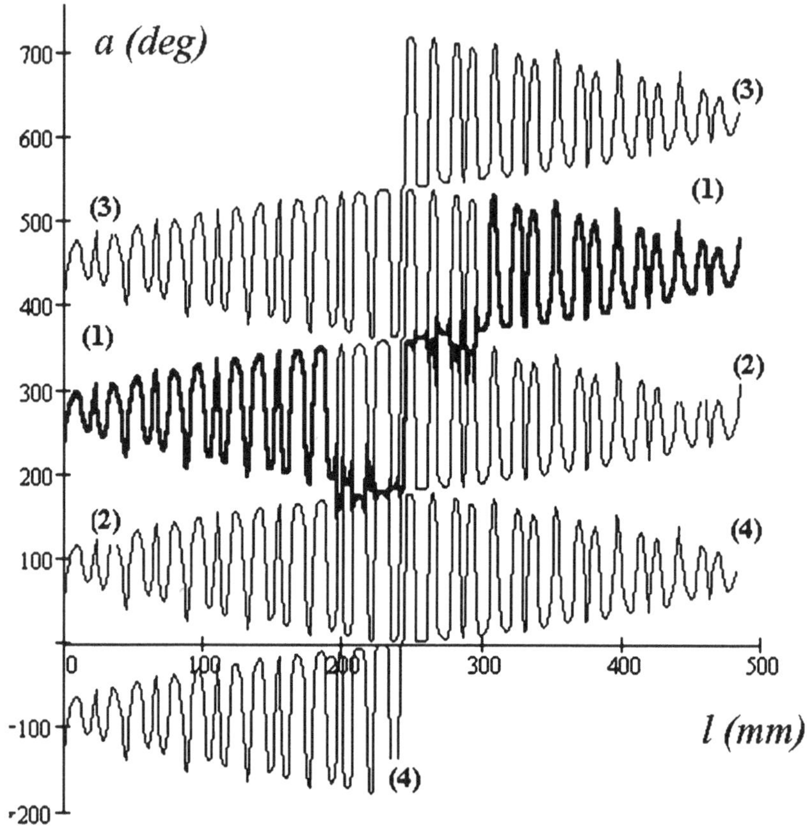

Fig. 6.10. Trajectories corresponding to (1) a_{base} (2) $a_{\text{base}} - \pi$ (3) $a_{\text{base}} + \pi$ (4) $a_{\text{base}} - 2\pi$ and the optimized trajectory (*bold line*) composed from the trajectories (1) (2) and (4)

Therefore, when the cutter location point is close to the center of the a-rotations, the tool should move "around the hill".

Given an average space step, one could evaluate the kinematics error for varying r_a. Such evaluation makes it possible to pre-compute the movements of the tool and to reduce the computational load spent on evaluating the error. Figure 6.12 shows the kinematics error corresponding to T_{across} and T_{around} as functions of a_2 and b_2 for $r_a = 0.1$ mm, 5 mm and 20 mm, respectively. Clearly, $r_a = 0.1$ mm requires T_{around} for any angle, $r_a = 20$ mm entails T_{across} whereas $r_a = 5$ mm requires a further evaluation of the kinematics error for each particular pair (a, b).

Such pre-computing performed for a set of prescribed spatial steps could constitute an optimization strategy for a practical five-axis machining. For instance, minimization of the kinematics error of the milling machine MAHO

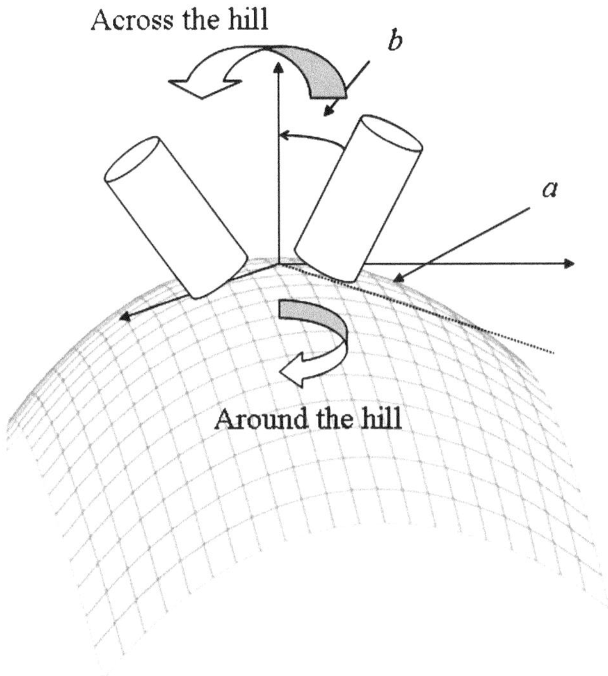

Fig. 6.11. Around or across the hill?

600E for $s = 1$ mm requires T_{around} if $r_a \in [0, 5]$ mm and T_{across} if $r_a > 10$ mm. However, when $r_a \in (5, 10)$ mm, the decision can not be pre-computed and shall be made by evaluating $\epsilon^{\text{kinematic}}$, explicitly.

Example 6.2. **Optimization with regard to the kinematics error**

We have shown that loops due to large variations in the rotation angles can be almost entirely eliminated for a trajectory having a single stationary point. Table 6.1 shows the performance of the method for an entire surface S_1 (Fig. 6.1) machined on MAHO 600E.

The optimization was performed without discriminating between ϵ_U and ϵ_O. Consider Table 6.1. Clearly, minimizing the total angle variation leads to minimizing the total kinematics error. On the contrary, an increase of the number of the CL points does not necessarily lead to a decrease of the error (see the lines 40×20 and 100×20) since the grids are not nested (the points of the coarse grid do not necessarily belong to the fine grid).

Example 6.3. **Optimization of the tool path with regard to overcut and undercut**

A very useful optimization can be performed when the undercuts and overcuts are differentiated. In order to detect them numerically every trajectory $W_{p,p+1}(t)$ is subdivided by a certain number of points t_k. At every point

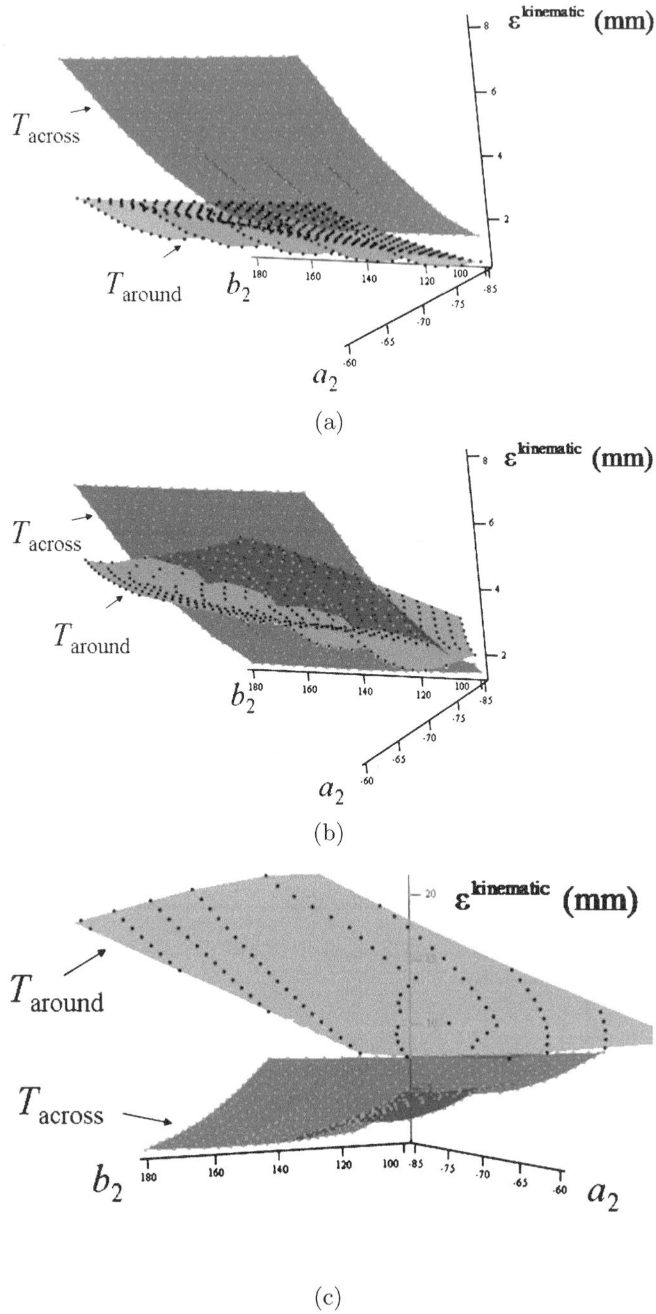

Fig. 6.12. The kinematic error as a function of a_2 and b_2 for rotation radius r_a = 0.1 mm (**a**), 5 mm (**b**) and 20 mm (**c**)

Table 6.1. Kinematics error for the optimized and non-optimized tool path, surface S_1 on MAHO 600E

	Max error (mm) /Max angle var (deg)			Path length (mm)
Grid size	No optimization	Optimization	Decrease (%)	Non-opt/Opt
10×20	23.862 / 168.779	12.426 / 78.072	47.925 / 53.743	2825.67 / 2255.54
15×20	19.300 / 162.186	8.517 / 101.524	55.870 / 34.403	2500.30 / 2123.02
20×20	20.228 / 160.080	7.558 / 84.927	62.636 / 46.947	2367.57 / 2101.15
30×20	16.253 / 141.890	7.162 / 88.230	55.934 / 37.818	2183.63 / 2038.27
40×20	8.711 / 103.885	6.878 / 88.399	21.042 / 14.907	2069.32 / 2020.11
100×20	7.395 / 90.898	7.103 / 89.102	3.949 / 1.976	1916.11 / 1911.17
130×20	3.999 / 68.416	3.999 / 68.416	0.000 / 0.000	1876.49 / 1876.49

$W_{p,p+1,k} = (x_{p,p+1,k}, y_{p,p+1,k}, z_{p,p+1,k})$ we find the corresponding parametric coordinates (u_k, v_k) by solving numerically the system

$$\begin{aligned} x_{p,p+1,k} &= x_S(u,v) \\ y_{p,p+1,k} &= y_S(u,v) \end{aligned} \tag{6.15}$$

Note that for explicit surfaces the above equations are linear and, therefore, a numerical solution is not required. Now, given the solution (u_k, v_k), we calculate $z_S(u_k, v_k)$ for every k. Next, we calculate the angle variations $c_{U,p,p+1}$ and $c_{O,p,p+1}$ for every p and for every graph node. Finally, we use the weights w_O and w_U and apply the Dijkstra's shortest path algorithm (see, for instance, [12]).

Consider machining surface S_2 on HERMLE UWF902H. Let $w_O = 1$, $w_U = 0$ that is, the minimization is performed only with regard to the overcuts. Figure 6.13a displays the conventional tool path with overcuts (the thick curves). Figure 6.13b shows how minimization of the total error eliminates the loops located at the right side of the workpiece. However, minimization of only the overcut error produces an entirely different path. The overcut loops at the boundary of the workpiece have been replaced by large but harmless undercuts (see Fig. 6.13c). Table 6.2 shows the behavior of the undercut and overcut error along with the angle variation as the number of points in the cutting direction increases. The decrease in the overcut error has been achieved at the expense of the increase of the undercuts and the length of the path.

Let us now demonstrate the techniques by optimizing the overcuts on MAHO 600E. The configuration of this machine will create an entirely different set of trajectories. The conventional tool path and the tool path optimized by means $w_O = 1$, $w_U = 1$ are displayed in Figs. 6.14(a-b), whereas a tool path optimized with $w_O = 1$, $w_U = 0$ in Fig. 6.14c. The corresponding machined surfaces are displayed in Figs. 6.15(a-c). Clearly, optimization with regard to only the overcuts makes it possible to cut a rough but reasonable surface which could be further refined whereas cutting without the overcut optimization may destroy the workpiece. The behavior of the kinematics error versus the number of points and the maximum angle variation is displayed in Table 6.3. Comparing HERMLE UWF902H and MAHO 600E (Tables 6.2 and 6.3) shows that the accuracy of the two cuts without optimization is comparable, whereas the optimized cut on MAHO 600E outperforms HERMLE UWF902H.

6.5 Uniform Angular Grids

This section introduces uniform angular grids inserted locally in such a way that the resulting total angle variation is minimized or at least reduced. Many algorithms to calculate the so-called feedrate (the step between the consecutive CL points) have been proposed. The simplest approach is to linearize the

Table 6.2. Overcut error optimization, surface S_2 on HERMLE UWF902H

	Max error (mm) / Max angle var (deg)				Path length (mm)	Angular variation (deg)
	No optimization		Optimization			
Grid size	Overcut	Undercut	Overcut	Undercut	Non-opt/Opt	Non-opt/Opt
10×20	19.11 / 139.78	23.40 / 169.95	13.33 / 109.90	56.16 / 356.22	2929.02 / 3641.59	7763.87 / 9317.35
15×20	9.54 / 32.96	19.28 / 162.13	9.52 / 32.76	60.93 / 364.84	2676.55 / 3985.66	7921.93 / 10131.55
20×20	8.07 / 88.17	20.21 / 160.01	4.41 / 23.83	60.93 / 364.84	2486.07 / 5021.87	7972.65 / 11477.04
30×20	6.28 / 149.23	16.24 / 141.87	2.40 / 45.76	51.55 / 372.37	2253.76 / 3845.65	8018.01 / 10119.48
40×20	5.41 / 139.58	8.81/ 104.05	1.60 / 14.187	13.24 / 130.02	2117.19 / 2166.59	8035.95 / 7993.36
100×20	0.00 / 4.00	7.40 / 90.83	0.00 / 4.00	7.13 / 89.27	1924.58 / 1919.35	8055.03 / 8032.72
130×20	0.00 / 3.09	4.00 / 68.34	0.00 / 3.09	4.00 / 68.34	1880.72 / 1880.72	8056.50 / 8056.50

Table 6.3. Overcut error optimization, surface S_2 on MAHO 600E

	Max error (mm) / Max angle var (deg)				Path length (mm)	Angular variation (deg)
	No optimization		Optimization			
Grid size	Overcut	Undercut	Overcut	Undercut	Non-opt/Opt	Non-opt/Opt
10×20	8.41 / 169.66	23.86 / 170.03	1.84 / 28.26	38.46 / 157.94	2825.67 / 3575.11	7768.03 / 7867.60
15×20	7.23 / 164.57	19.30 / 162.20	1.98 / 40.47	32.92 / 159.15	2500.30 / 2417.30	7926.18 / 7339.96
20×20	6.78 / 159.42	20.23 / 160.83	0.96 / 26.98	7.56 / 87.37	2367.57 / 2101.14	7976.88 / 7148.41
30×20	6.25 / 149.31	16.25 / 141.94	0.00 / 12.57	7.16 / 89.19	2183.63 / 2032.20	8022.24 / 7848.15
40×20	5.39 / 139.65	8.71 / 104.12	0.00 / 9.64	6.88 / 88.56	2069.32 / 2020.11	8040.19 / 7788.45
100×20	0.00 / 4.00	7.40 / 90.90	0.00 / 4.00	7.10 / 89.20	1916.11 / 1911.17	8059.30 / 8036.42
130×20	0.00 / 3.09	4.00 / 68.42	0.00 / 3.09	4.00 / 68.41	1876.49 / 1876.49	8060.76 / 8080.76

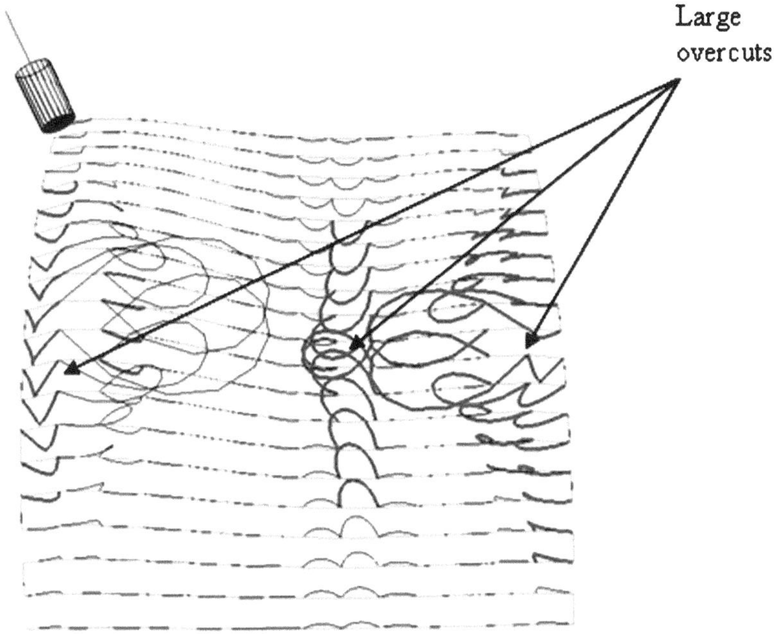

Fig. 6.13a. Conventional tool path for S_2 on HERMLE UWF902H showing large overcuts (*thick curves*)

surface along the tool path, calculate the error and insert additional points until the error is within the required tolerance (see [5]). Another approach is to interpolate the desired trajectory by a spline or a NURBS and establish the required feedrate using properties of the curve. The feedrate may depend on the space coordinates of the curve as well as on the angular changes (see, for example, [13]). A combination of the actual kinematics of the machine with an interpolating method is made by Lo [6].

A general framework for applying grid generation to tool path optimization proposed in [7] and in [8] is given in Chap. 5. In particular, grid generation based on the Dirichlet functional, invoked in an iterative loop with a suitable preprocessing, may substantially decrease the kinematics error. The grid generation is applicable if $\epsilon \to 0$ as the area of the grid cell A tends to zero. If it is the case, a functional representing the so-called equi-distribution principle [8] and subjected to the machine constraints can be introduced.

Now we will consider a different situation characterized by a rough cut, when the tool rotations angles jump as described above. In this case the adaptive grid may require very small spatial cells. As a result the algorithm may create twisted and degenerated cells which may not converge or display a very slow convergence.

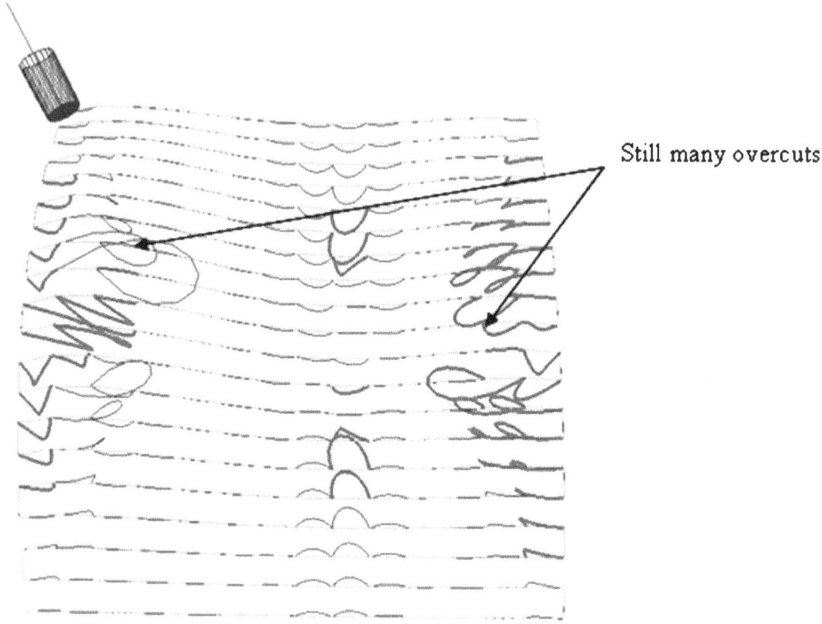

Fig. 6.13b. The tool path optimized with regard to the total error. There are still many trajectories with large overcuts

Suppose we have constructed a "basic" tool path represented by a rectangular or a curvilinear grid described in Chap. 5. Suppose we are allowed to "inject" a certain number of additional points to decrease the error. Where and how should these points be inserted? Let us employ local grids with equal angular increments around points characterized by large angle variations. We shall call such grids the uniform angular grids.

Recall that near the stationary points there often exists a neighborhood such that one of the rotations is not performed or one of the angles changes faster than another angle. In this case it is sufficient to construct a grid uniform only with regard to the "fast" angle.

The algorithm consists of the following steps. First of all, it detects points characterized by sharp angular variation. Next, for each point the algorithm determines the positions between which the uniform angular grid should be constructed. Usually the interval includes a few tool positions right before and after the singularity which creates the kinematics loop. The space between every pair of the selected points W_p and W_{p+1} must be subdivided in such a way that $\Delta a_{p,p+1,i,i+1} \approx \text{const}$ or $\Delta b_{p,p+1,i,i+1} \approx \text{const}$ for every subinterval $[i, i+1]$.

We have already noted that the stationary point is cut by either changing sharply the first or the second rotation angle. Irrespectively whether the singular point is a CC point or not, the tool will follow a path "across or

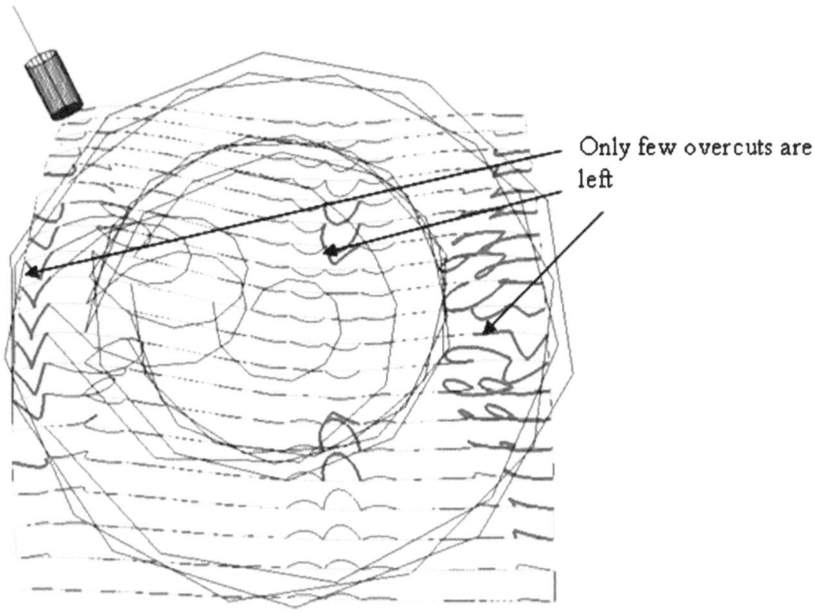

Fig. 6.13c. The tool path optimized with regard to the overcut error. The large overcuts have been replaced by large but harmless undercuts

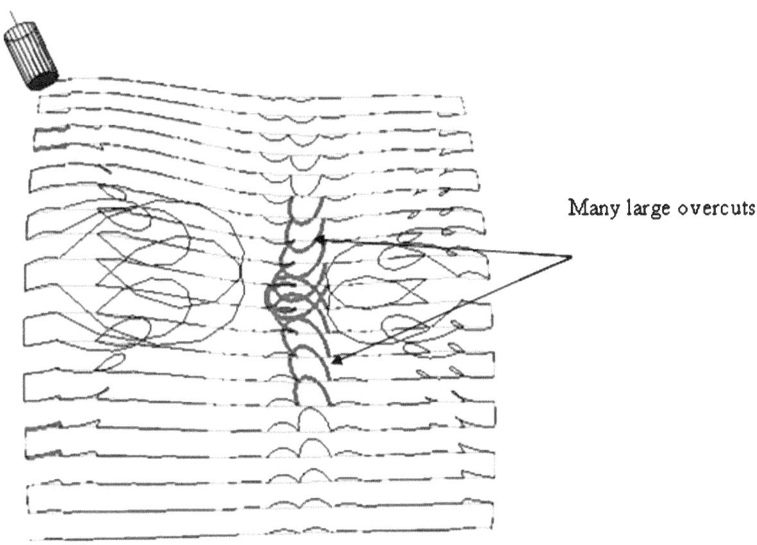

Fig. 6.14a. Conventional tool path for S_2 on MAHO 600E

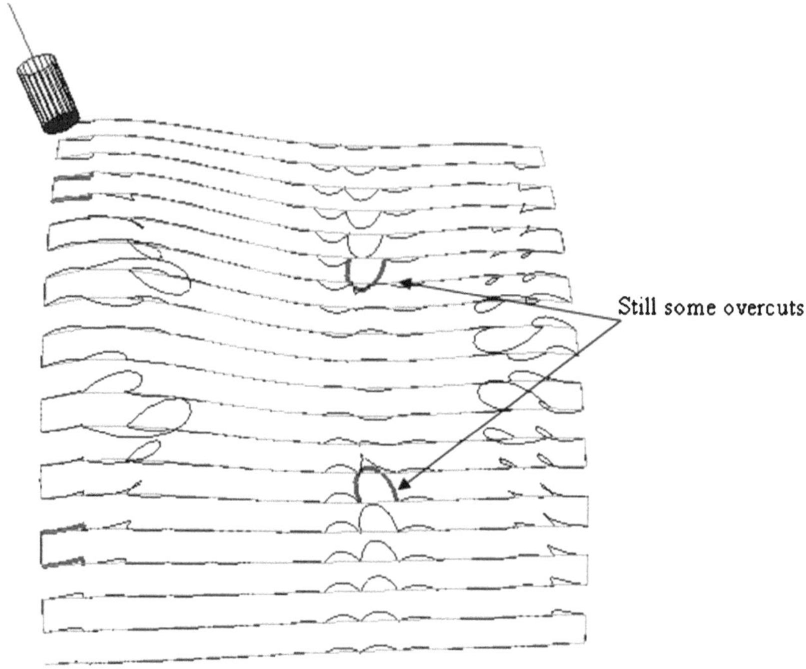

Fig. 6.14b. Optimization with regard to the total error

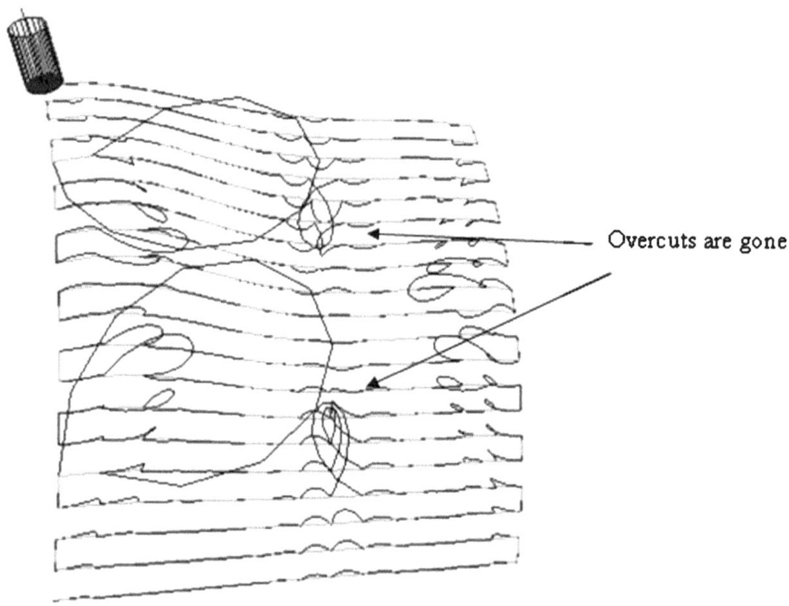

Fig. 6.14c. Optimization with regard to the overcut error, S_2 on MAHO 600E

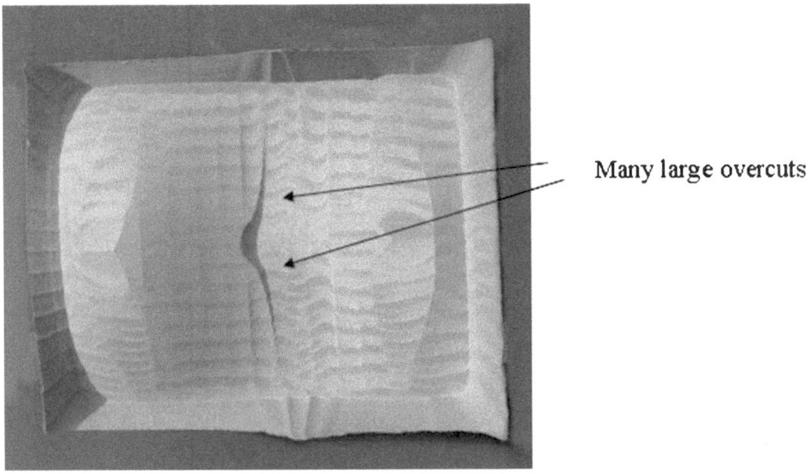

Fig. 6.15a. Without optimization, S_2 on MAHO 600E (corresponds to Fig. 6.14a)

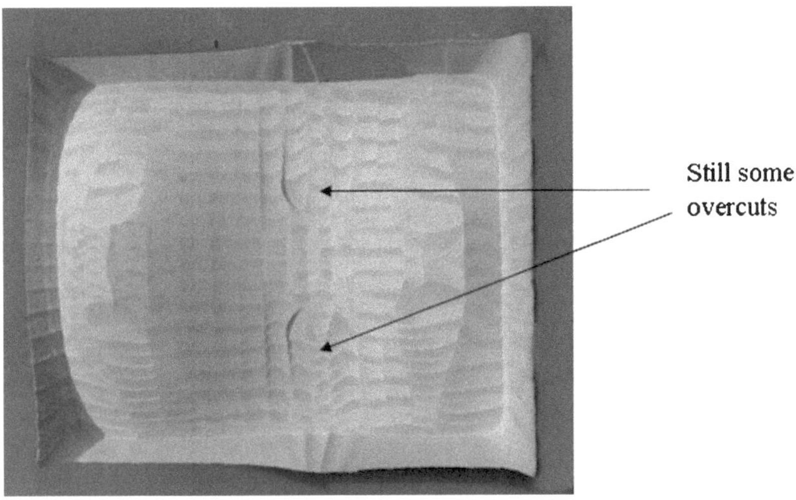

Fig. 6.15b. Optimization with regard to the total error, S_2 on MAHO 600E (corresponds to Fig. 6.14b)

around the hill" depending on the choice of the rotation angles. Consequently, the algorithm selects the appropriate angle (say, angle a) and subdivides the angular interval $\Delta a_{p,p+1}$ into equal subintervals $\Delta a_{p,p+1,i,i+1}$ such that $a_{p,p+1,i} = a_p + \Delta_{p,p+1} i$, where $\Delta_{p,p+1}$ is the corresponding step. The corresponding spatial positions s_i are found from the surface equation as follows. First of all, we specify a parametric spatial trajectory $T = T(s)$ on the surface through the stationary point. Next, for each i we solve numerically the

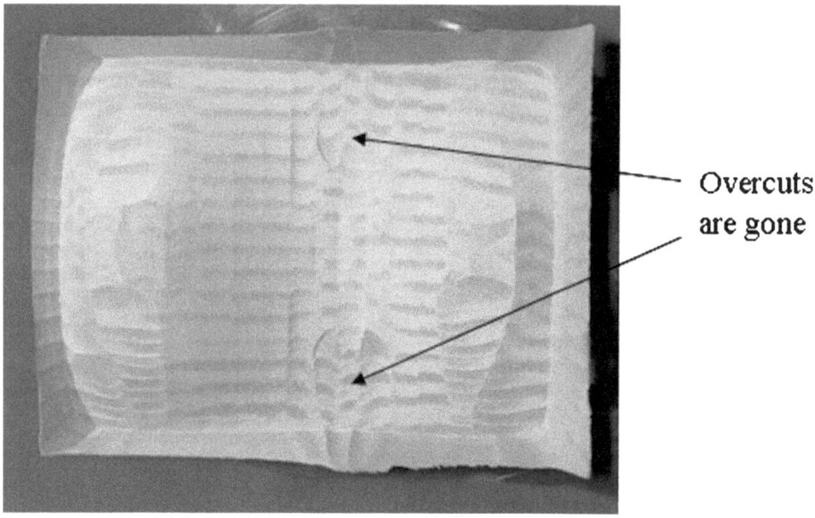

Fig. 6.15c. Optimization with regard to the overcut error, S_2 on MAHO 600E (corresponds to Fig. 6.14c)

equation $a(T(s)) = a_{p,p+1}$, where $a(T(s))$ denotes the rotation angle on $T(s)$ calculated from the surface normal.

6.6 Uniform Angular Grids: Numerical and Machining Experiments

Consider machining surface S_2 by MAHO 600E. We are allowed to "inject" additional points in order to decrease the error. First of all, consider the surface machined using the isoparametric tool path represented by a local uniform grid (see Fig. 6.14a). The large error loops appearing due to sharp angle variations can be eliminated by inserting the additional points in these areas using a uniform angular distribution of the points shown in Fig. 6.16b. The substantial reduction of the error (50 times!) has been achieved by inserting only 8 points inside each loop characterized by large error.

It may seem that inserting an equi-spaced set of points may lead to the same or similar result. However, it is not the case. The uniform spatial distribution of the points does not have a significant impact on the error. The error has been reduced only by 4 times by inserting the 8 additional points. Table 6.4 displays the kinematics errors and the rotation angles before and after applying the proposed point insertion. Clearly, the uniform angular grid allows to substantially decrease the error, whereas the conventional is not efficient.

However, when a large number of the additional points have been inserted, the error decrease is approximately the same for the space or angular inserting. Therefore, the method applies to rough cuts characterized by sharp variations

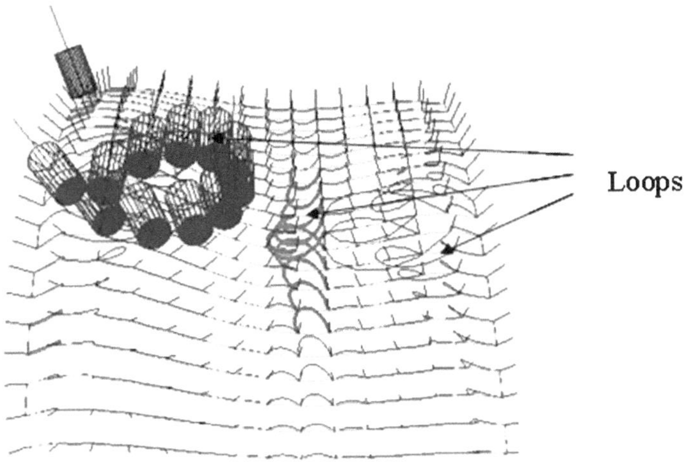

Fig. 6.16a. Tool path and tool orientations, S_2 on MAHO 600E, before insertion

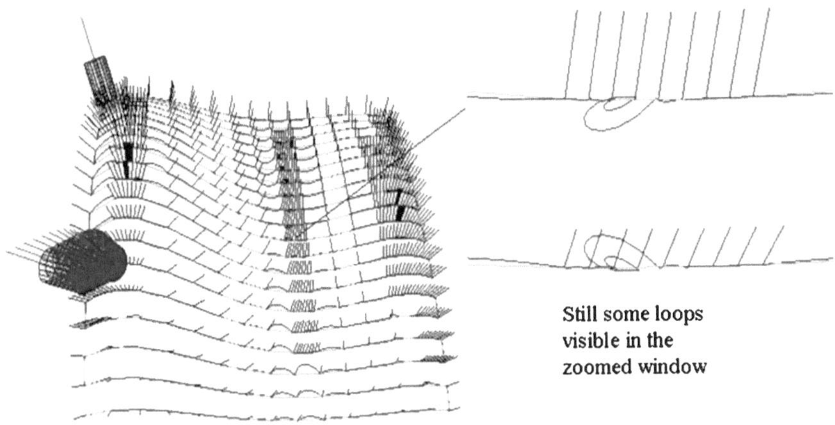

Fig. 6.16b. Tool path and tool orientations, S_2 on MAHO 600E, conventional point insertion

Table 6.4. Error versus number of inserted points, the basic grid size 15×20

Number of inserted points	Max error (mm) Conventional / Angular grid	Angular variation (deg) Conventional / Uniform angular grid
0	19.300 / 19.300	162.202 / 162.202
8	4.241 / 0.416	68.660 / 20.631
16	1.707 / 0.138	43.016 / 10.629
32	0.490 / 0.092	20.647 / 5.352
64	0.158 / 0.089	10.955 / 2.680
128	0.099 / 0.089	5.378 / 1.340

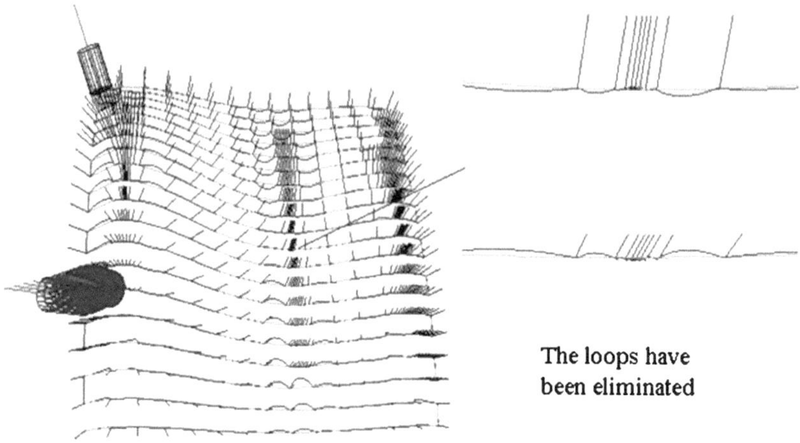

Fig. 6.16c. Tool path and tool orientations, S_2 on MAHO 600E, angular grid insertion

of the rotation angles. Figures 6.16(b-c) explain advantages of the angular grids. Clearly, the conventional uniform grid does not remove the kinematics loops although the amplitude has been decreased. As opposed to that, the angular grid changes the behavior of the trajectories by entirely eliminating the loops. Figures 6.17(a-b) show the difference between the workpieces machined using the conventional and the angular grid technologies.

The cutting experiments have been performed with the flat-end cutter selected due to its popularity for fast, rough milling when the sharp angular variations are the most expected. However, the angular grids are applicable to the ball nose cutter being often used for finish machining. As a matter of fact, the algorithms apply to a general APT cutter (Automatic Programmed Tool) without major modifications. The equations of the APT cutter include the most popular shapes such as the flat-end shaped cutter, toroidal cutter, the ball nose cutter (see the Appendix).

Combination of the two proposed methods might lead to even better optimizations. However usually, the angular grid should be constructed before the shortest path procedure or in an iterative loop presented below. Even when the angular grid is applicable after the shortest path routine, the implementation is not as straightforward as it may seem since constructing the angular grid after the shortest path optimization requires to keep track of changes associated with every point. Furthermore, applying the angular grid after the optimal sequencing is often not possible because inserting even one point may create an entirely different graph for the shortest path optimization.

Let us illustrate this case by an example. Consider surface S_2 with the tool path obtained after the optimal sequencing (see Fig. 6.18). The largest loop between point 141 and 142, which can not be treated by the optimal sequencing, is on the right side of the surface. The angles before applying the algorithm are $a_{141} = 319°$, $b_{141} = -79°$, $a_{142} = 224°$, $b_{142} = -80°$. The shortest path optimization produces $a_{141} = 319°$, $b_{141} = -79°$, $a_{142,\text{new}} = a_{142} + 180° = 404°$, $b_{142,\text{new}} = -b_{142} - 180° = -100°$. Inserting a point in the middle of the loop yields $a_{\text{mid}} = 267°$, $b_{\text{mid}} = -82°$.

Taking into account the "history", we modify the pair of angles as follows $a_{\text{mid}} = a_{\text{mid}} + 180° = 447°$, $b_{\text{mid}} = -b_{\text{mid}} - 180° = -98°$ which leads to a larger loop depicted in Fig. 6.19. Note that $a_{\text{mid}} \notin [a_{141}, a_{142}]$ anymore. In other words, the additional point has destroyed the particular shortest path. Since the remaining part of the shortest path depends on these angles, the entire path has been destroyed and the optimization should be performed again.

The following procedure works very well when combining the two methods. First, we apply the optimal sequencing. Second, we select the largest loops appearing after applying the optimal sequencing. Next, we go back to the original angles and insert angular grids into the selected loops. Finally, we apply the optimal sequencing again to the modified tool path.

If the required tolerance has not been achieved, or even some new loops have appeared, we select the largest loops again and go back to inserting points into the original tool path.

Finally, apart from the machine kinematics, the accuracy of machining is often affected by the tool accelerations and decelerations due to frequent changes of the tool path directions. Since the phenomenon often occurs in the areas of sharp variations of the rotation angles, let us discuss the proposed method with the reference to the acceleration errors.

In order to reduce the acceleration error the entire tool path is usually treated as an interpolating curve parameterized with regard to the chord length between two consecutive reference points. Generating the tool positions by incrementing the chord length leads to the feedrate instabilities due to the difference between the chord and the arc length. The instabilities induce undesirable accelerations and jerk fluctuations (see our literature survey on interpolators in Chap. 1).

Our procedures apply to the regions where the angular variations are the most source of errors. However, the procedures do not treat the accelerations explicitly. Therefore, when such accelerations lead to large inaccuracies, the proposed algorithms should be combined with the above mentioned spline interpolating methods. However, observe that reducing the *angular* variation leads to equal increments in the angular space. Therefore, it reduces *angular accelerations* appearing in the five-axis mode due to sharp variations of the tool orientations.

Ideally, in five-axis machining the tool path must be regarded as a curve in the five dimensional space and parameterized with regard to its arc length in 5D. Such parameterizations should be associated with several sources of error such as the kinematics error, errors due to accelerations, etc. Unfortunately, such parameterizations still constitute an open problem.

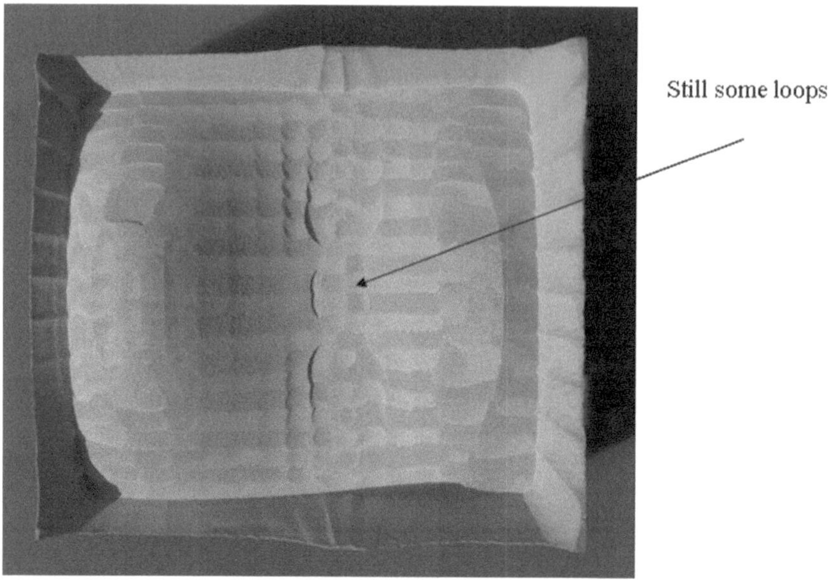

Fig. 6.17a. Spatial grid for S_2 on MAHO 600E (corresponds to Fig. 6.16b)

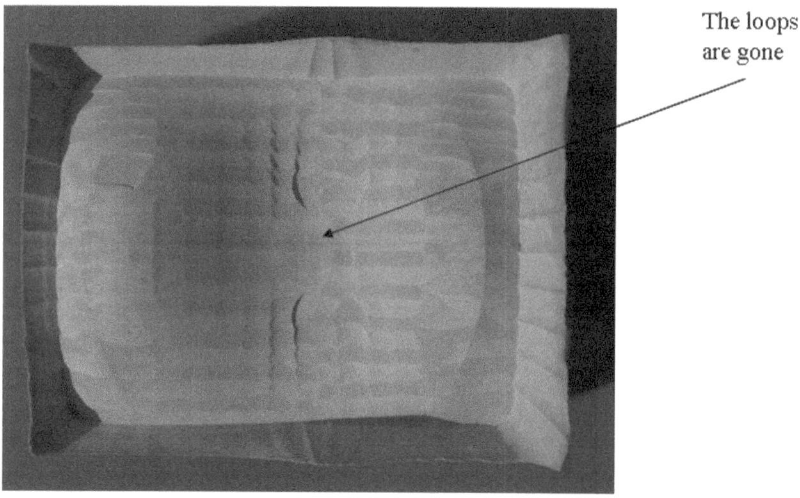

Fig. 6.17b. Angular grid insertion for S_2 on MAHO 600E (corresponds to Fig. 6.16c)

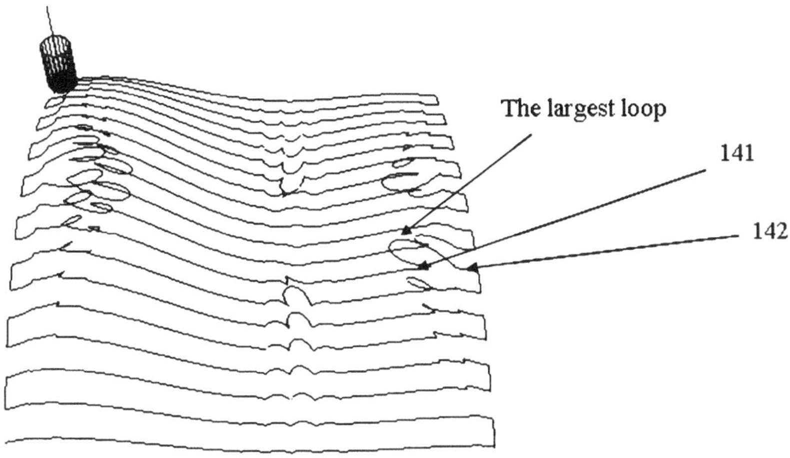

Fig. 6.18. Tool path for surface S_2 on MAHO 600E after the optimal sequencing

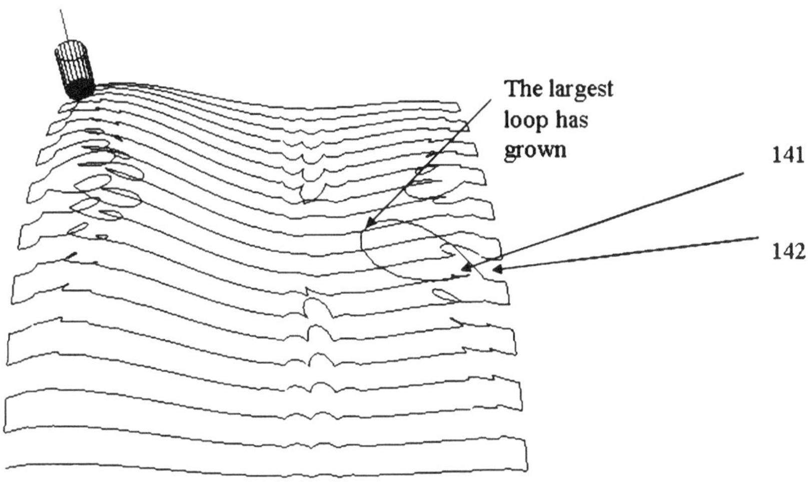

Fig. 6.19. Tool path for surface S_2 on MAHO 600E after the optimal sequencing and inserting one point into the largest loop

Appendix: The APT cutter

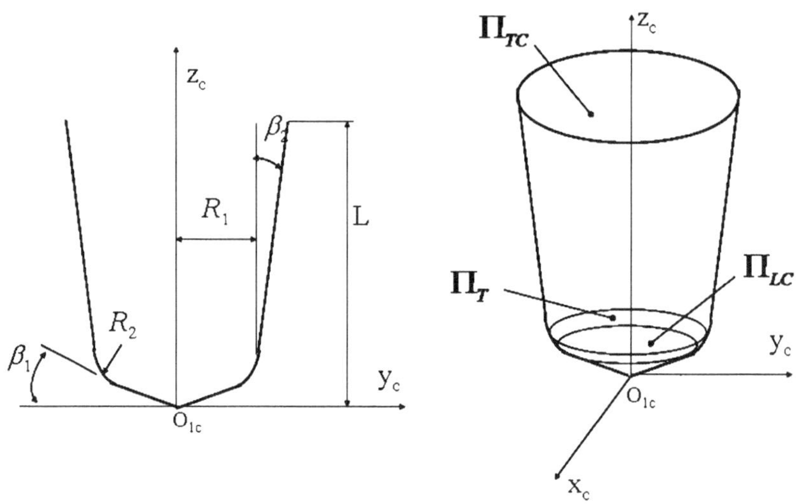

Fig. 6.20. APT cutter geometry

The cutter surface is composed of the truncated cone, torus and lower cone:

$$\Pi_C = \Pi_{TC} \cup \Pi_T \cup \Pi_{LC},$$

where

$$\Pi_{TC} = \begin{bmatrix} (R_1 + t\tan\beta_2)\sin\gamma \\ (R_1 + t\tan\beta_2)\cos\gamma \\ t + h - R_2\sin\beta_2 \end{bmatrix},$$

$\gamma \in [0, 2\pi]$, $t \in [h - R_2\sin\beta_2, L]$,

$h = R_2\cos\beta_1 + R_2\cos\beta_1\tan\beta_1 + R_1\tan\beta_1 - R_2\tan\beta_1\cos\beta_2$,

$$\Pi_T = \begin{Bmatrix} x_c \\ y_c \\ z_c \end{Bmatrix} = \begin{bmatrix} (e + R_2\sin\chi)\sin\gamma \\ (e + R_2\sin\chi)\cos\gamma \\ h - R_2\cos\chi \end{bmatrix},$$

$\chi \in [\beta_1, \frac{\pi}{2} - \beta_2]$, $e = R_1 - R_2\cos\beta_2$,

$$\Pi_{LC} = \begin{Bmatrix} x_c \\ y_c \\ z_c \end{Bmatrix} = \begin{bmatrix} t\cot\beta_1\sin\gamma \\ t\cot\beta_1\cos\gamma \\ t \end{bmatrix},$$

$t \in [0, h - R_2\cos\beta_1]$.

In the case of specific fillet-end cutter, $\beta_1 = \beta_2 = 0$. The flat-end cutter is a particular case of the fillet-end cutter when $R_2 = 0$.

References

[1] Affouard, A., Duc, E., Lartigue, C., Langeron, J.-M., and Bourdet, P. 2004. Avoiding 5-axis singularities using tool path deformation. *International Journal of Machine Tools and Manufacture*, 44(4):415–425.

[2] Arkin, E. M., Bender, M. A., Demaine, E. D., Fekete, S. P., Mitchell, J. S. B., and Sethia, S. 2001. Optimal covering tours with turn costs. In *Proceedings of the 12th Annual ACM-SIAM Symposium on Discrete Algorithms (SODA 2001)*, pages 138–147.

[3] Arkin, E. M., Bender, M. A., Demaine, E. D., Fekete, S. P., Mitchell, J. S. B., and Sethia, S. 2005. Optimal covering tours with turn costs. *SIAM Journal on Computing*, 35(3):531–566.

[4] Jung, Y. H., Lee, D. W., Kim, J. S., and Mok, H. S. 2002. NC postprocessor for 5-axis milling machine of table-rotating/tilting type. *Journal of Materials Processing Technology*, 130-131:641–646.

[5] Li, F., Wang, X. C., Ghosh, S. K., Kong, D. Z., Lai, T. Q., and Wu, X. T. 1995. Tool-path generation for machining sculptured surface. *Journal of Materials Processing Technology*, 48(1):811–816.

[6] Lo, C. C. 1999. Efficient cutter-path planning for five-axis surface machining with a flat-end cutter. *Computer-Aided Design*, 31(9):557–566.

[7] Makhanov, S. S., Batanov, D., Bohez, E., Sonthipaumpoon, K., Anotaipaiboon, W., and Tabucanon, M. 2002. On the tool-path optimization of a milling robot. *Computers & Industrial Engineering*, 43(3):455–472.

[8] Makhanov, S. S. and Ivanenko, S. A. 2003. Grid generation as applied to optimize cutting operations of the five-axis milling machine. *Applied Numerical Mathematics*, 46(3-4):331–351.

[9] Makhanov, S. S. and Munlin, M. Optimal sequencing of rotation angles for five-axis machining. *International Journal of Advanced Manufacturing Technology*. in press.

[10] Marchuk, G. I. 1982. *Methods of Numerical Mathematics*. Springer, New York.

[11] Munlin, M.-A., Makhanov, S. S., and Bohez, E. L. J. 2004. Optimization of rotations of a five-axis milling machine near stationary points. *Computer-Aided Design*, 36(12):1117–1128.

[12] Weiss, M. A. 1993. *Data structures and algorithm analysis in C*. Benjamin-Cummings Publishing Co., Inc., Redwood City, CA, USA.

[13] Xu, H.-Y. 2003. Linear and angular feedrate interpolation for planar implicit curves. *Computer-Aided Design*, 35(3):301–317.

7

Theory of Optimal Setup for Five-Axis NC Machining

7.1 Introduction

In five-axis machining, the forward step (kinematics) error due to nonlinearity of the machine kinematics can significantly affect the quality of the machined surfaces. In a simple approach, kinematics error between two CC points can be reduced to within a prescribed tolerance by inserting enough points. However, this method is often inefficient, especially for high speed milling where a tool path containing a large number of points can slow down the milling process and is, therefore, undesirable. A technique of angle variation presented in Chap. 6 can also be used to minimize the kinematics error without inserting additional points. The error after minimization using this angle adjustment method can still, however, exceed the prescribed tolerance and more points or another minimization techniques are required.

In can be shown that an initial workpiece setup as well as parameters related to the machine configuration such as the position of each center of rotation and the tool length can affect the machining error due to the kinematics of the machine. Analysis of an initial setup is not currently provided by commercial CAD/CAM software tools such as Unigraphics, EdgeCam, and Vericut. However, in the academic literature, a few researchers have addressed the subject. Jung et al. [3] propose a general five-axis postprocessor, but the possibility of optimization has not been raised. Sijie et al. [6] optimize the position and orientation of the stock relative to the CAD modeling surface but machine related parameters were not optimized. Surveys and analysis of five-axis configuration are given in [2, 7, 8]. However, optimizing machining setups for specific surfaces has apparently been overlooked.

This chapter presents a new optimization model designed to minimize kinematics error introduced by the initial setup of a five-axis milling machine. An initial setup consists of the position and orientation of the workpiece with respect to the mounting table and, optionally, the machine's initial configuration. Depending on the type of the machine (see machine classification in Sect. 2.3), some parameters have no effect on the kinematics of the machine.

These parameters are termed invariant parameters (see Sect. 7.2.1). Furthermore, for each type of machine, there exists variables that are dependent and only one of these need to be optimized (see Sect. 7.3.2). After identifying the sets of invariant and dependent variables, a corresponding system of nonlinear equations is constructed and solved numerically.

For demonstration propose, the accuracy achieved through use of the proposed optimization of the initial setup is compared with the accuracy achieved through use of a "standard" positioning of the workpiece where a blank part is placed so that the bottom of the machined part is at a certain distance above the mounting table to ensure against collision avoidance. The center of the base is positioned at the center of the mounting table, and the **z**-axis in workpiece coordinates is perpendicular to the mounting table.

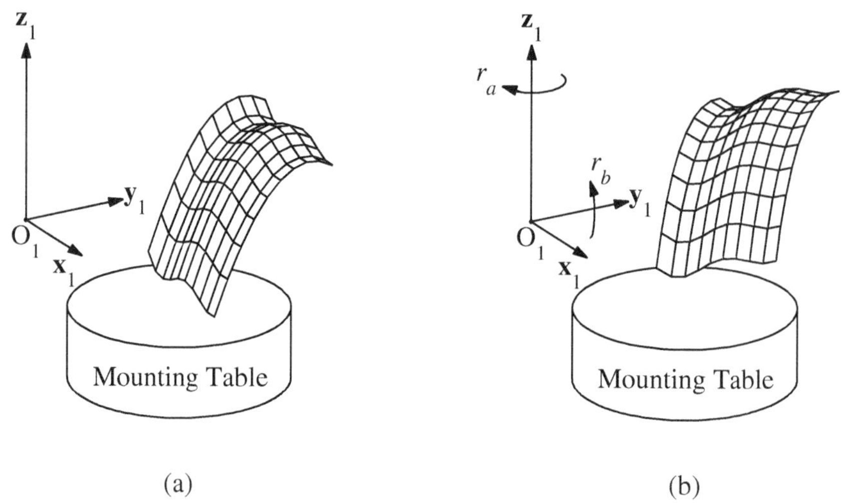

Fig. 7.1. Changing an initial workpiece setup by rotating around the $\mathbf{z}_1$-axis and the $\mathbf{y}_1$-axis

The following introductory example shows that optimization can be surprisingly effective. Let W_1 and W_2 be two successive spatial positions of the tool tip (CC points) in workpiece coordinates and let I_1 and I_2 be the corresponding tool orientations. The actual tool trajectory between W_1 and W_2 is generally nonlinear due to nonlinearity of the machine's kinematics; it also depends on the initial workpiece position and orientation with respect to the mounting table. Figure 7.1 shows two different initial workpiece setups. The workpiece orientation in Fig. 7.1(b) is obtained by first rotating the workpiece in Fig. 7.1(a) around the $\mathbf{z}_1$-axis by r_a and then rotating it around the $\mathbf{y}_1$-axis by r_b. Finally, it is shifted by T_{12}.

The new workpiece setup changes the cutter contact points and the orientations of the tool as follows:

$$
\begin{aligned}
W'_p &= R_b[r_b]R_a[r_a]W_p + T_{12}, \\
I'_p &= R_b[r_b]R_a[r_a]I_p,
\end{aligned}
\tag{7.1}
$$

where R_a and R_b are the rotation matrices corresponding to r_a and r_b. Figure 7.2 illustrates the effect of workpiece setup on the tool's actual trajectory. Astonishingly, with the setup shown in Fig. 7.2(c), the error between the desired tool trajectory $W_{12}^D(t)$ and the actual tool trajectory $W_{12}(t)$ has been reduced by a factor of one thousand! (the error in the optimal setup is 0.03 mm). This error reduction cannot be achieved by using the method of angle adjustment alone (see Sect. 3.4 and Fig. 3.8 on Pages 63-67).

In general, the optimization parameters $\mathfrak{M}$ are (see Figs. 2.4-2.6):

1. The workpiece setup $\mathfrak{M}_S = \{T_{12}, r_a, r_b\}$, where $T_{12} = (T_{12,x}, T_{12,y}, T_{12,z})$ is the position of the origin of the workpiece coordinates with respect to O_2 (the coordinate system of the first rotary part) and r_a and r_b are the angles of rotation corresponding to the initial orientation of the workpiece.
2. The machine design $\mathfrak{M}_T = \{T_{23}, T_{34}, T_4\}$, where $T_{23} = (T_{23,x}, T_{23,y}, T_{23,z})$ denotes the position of the origin of the first rotary part in the coordinate system of the second rotary part, $T_{34} = (T_{34,x}, T_{34,y}, T_{34,z})$ denotes the position of the second rotary part in the spindle coordinate system, and T_4 is a vector describing the tool length. Note that the tool length L is being treated as a translation T_4 which is either $(0, 0, L)$ or $(0, 0, -L)$ depending on the position of the tool tip in the spindle coordinate system.

Consider the following optimization problem:

$$
\min_{\mathfrak{M}} \ \epsilon,
\tag{7.2}
$$

where ϵ is the kinematics error defined as the difference between the desired and the actual trajectory of the tool tip. $\mathfrak{M} = \{\mathfrak{M}_S, \mathfrak{M}_T\}$ consists of the twelve parameters introduced above. It will be proven that only six parameters from the twelve need to be considered, since the remaining six parameters do not have any effect on the kinematics error. The actual identities of the six parameters to be ignored depend on the type of five-axis milling machine being used (see Sect. 2.3). The six parameters are selected as follows. First, for each type of machine, some of the setup parameters have no effect on the tool trajectory at all. These parameters can be immediately excluded from the optimization. Next, for each type of machine, an additional set of dependent parameters is found. Although each of these parameters does affect the tool trajectory when manipulated independently, each can be replaced by a linear combination of the other parameters, so these linearly dependent parameters can also be eliminated. Finally, the elimination of invariant and linearly dependent variables leads to a nonlinear system of equations that can be solved numerically by a Newton-Raphson procedure.

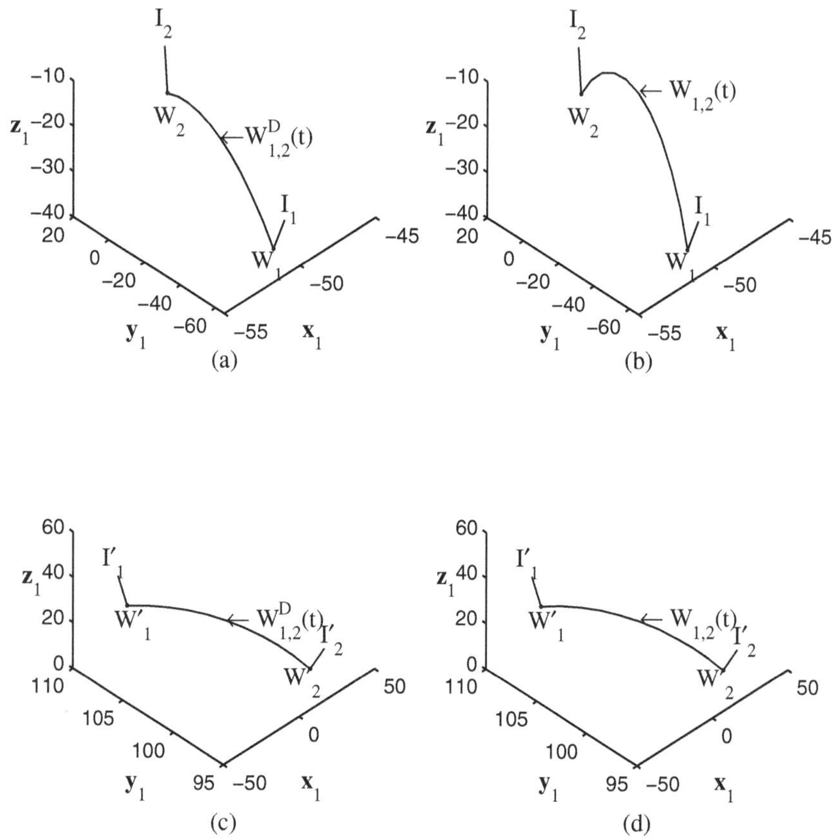

Fig. 7.2. The desired tool trajectory (**a**) and the actual tool trajectory before optimization (**b**). The desired tool trajectory (**c**) and the actual tool trajectory for the new (optimal) setup (**d**) (the workpiece is rotated by 99.23° around $\mathbf{z}_1$-axis and −41.28° around the $\mathbf{y}_1$-axis and shifted by [21.47 49.33 33.67] mm). The desired and actual tool trajectories almost coincide in the optimal setup. The error has been reduced by 99.9%

The solution of this system establishes, for a particular machine and desired surface, not only the optimal position and orientation of the workpiece with respect to the mounting table, but also the optimal initial configuration of the five-axis machine. As previously mentioned, the method has proven extremely effective in many cases. Moreover, it requires nothing of the operator except changing the initial setup, which is a very simple, nearly zero-cost, operation.

7.2 Tool Trajectory Analysis

This section presents a theoretical analysis of the tool tip trajectory for each of the three types of five-axis machines. First, parameters that do not affect the tool trajectory are identified. These parameters will be excluded from the optimization procedure. Next, the effect of initial workpiece setup on tool trajectory is described.

7.2.1 Invariant Parameters

Five-axis machines are characterized by nonlinear kinematics due to the two additional degrees of freedom that control tool orientation. In general, the trajectory of the tool tip is nonlinear. The closed-form representation of the tool tip trajectory for general machine kinematics is given in (3.48). In each of the three types of five-axis machines (see Sect. 2.3), there are some parameters that have no effect on the tool trajectory. These parameters are identified by the following theorems.

Theorem 7.1. *The kinematics of the 2-0 machine implies that the tool trajectory between two arbitrary points is invariant with respect to T^*_{34} and $T_{23,y}$.*

Proof. Denote $A[a(t)]$ by A, $B[b(t)]$ by B, $A[a_p]$ by A_p, $B[b_p]$ by B_p. Represent $\mathfrak{K}^{-1}(W)$ by $\mathfrak{K}^{-1}(W) = GB\hat{\mathfrak{K}}^{-1}(W)+T^*_{34}$, where $\hat{\mathfrak{K}}^{-1}(W) = A(W+T_{12})+T_{23}$ (see (2.5)). The tool tip trajectory (3.47) is then given by

$$W_{p,p+1}(t) = \mathfrak{K}(tGB_{p+1}\hat{\mathfrak{K}}^{-1}_{p+1} + tT^*_{34} + (1-t)GB_p\hat{\mathfrak{K}}^{-1}_p + (1-t)T^*_{34}), \quad (7.3)$$

where $\hat{\mathfrak{K}}^{-1}_p$ denotes $\hat{\mathfrak{K}}^{-1}(\mathfrak{M}, \mathfrak{N}_p, W_p)$. Clearly, from (2.5), $W = \mathfrak{K}(M) = A^{-1}(B^{-1}G^{-1}(M-T^*_{34})-T_{23})-T_{12}$. Consider $W = \mathfrak{K}(M) = \hat{\mathfrak{K}}(S) = A^{-1}(S-T_{23}) - T_{12}$ where $S = B^{-1}G^{-1}(M - T^*_{34})$. Rewriting (7.3) in terms of $\hat{\mathfrak{K}}(S)$ yields

$$\begin{aligned} W_{p,p+1}(t) &= \hat{\mathfrak{K}}(B^{-1}G^{-1}(tGB_{p+1}\hat{\mathfrak{K}}^{-1}_{p+1} + tT^*_{34} \\ &\quad + (1-t)GB_p\hat{\mathfrak{K}}^{-1}_p + (1-t)T^*_{34} - T^*_{34})), \\ &= \hat{\mathfrak{K}}(tB^{-1}B_{p+1}\hat{\mathfrak{K}}^{-1}_{p+1} + (1-t)B^{-1}B_p\hat{\mathfrak{K}}^{-1}_p). \end{aligned} \quad (7.4)$$

Since neither $\hat{\mathfrak{K}}$ nor $\hat{\mathfrak{K}}^{-1}$ depends on T^*_{34}, $W_{p,p+1}(t)$ does not depend on T^*_{34} either. Therefore, the tool trajectory is invariant with respect to T^*_{34}.

Let us now prove the invariance with regard to $T_{23,y}$. Substituting $\hat{\mathfrak{K}}^{-1}(W) = A(W+T_{12})+T_{23}$ and $\hat{\mathfrak{K}}(S) = A^{-1}(S-T_{23})-T_{12}$ into (7.4) yields

$$\begin{aligned} W_{p,p+1}(t) &= \hat{\mathfrak{R}}(tB^{-1}B_{p+1}\hat{\mathfrak{R}}_{p+1}^{-1} + (1-t)B^{-1}B_p\hat{\mathfrak{R}}_p^{-1}), \\ &= \hat{\mathfrak{R}}(tB^{-1}B_{p+1}(A_{p+1}(W_{p+1}+T_{12})+T_{23}) \\ &\quad + (1-t)B^{-1}B_p(A_p(W_p+T_{12})+T_{23})), \\ &= A^{-1}(tB^{-1}B_{p+1}(A_{p+1}(W_{p+1}+T_{12})+T_{23}) \\ &\quad + (1-t)B^{-1}B_p(A_p(W_p+T_{12})+T_{23})-T_{23})-T_{12}, \\ &= A^{-1}(tB^{-1}B_{p+1}A_{p+1}(W_{p+1}+T_{12}) \\ &\quad + (1-t)B^{-1}B_pA_p(W_p+T_{12}) \\ &\quad + tB^{-1}B_{p+1}T_{23} + (1-t)B^{-1}B_pT_{23} - T_{23}) - T_{12}. \end{aligned} \tag{7.5}$$

The terms involving T_{23} in (7.5) can be written $(tB^{-1}B_{p+1}T_{23} + (1-t)B^{-1}B_pT_{23} - T_{23})$. Since B^{-1}, B_{p+1} and B_p are rotations around the **y**-axis, they do not change the y-component of T_{23}. Consequently, $(tB^{-1}B_{p+1}T_{23} + (1-t)B^{-1}B_pT_{23} - T_{23})_y = 0$. Therefore, the tool trajectory does not depend on $T_{23,y}$. This completes the proof of the Theorem.

Theorem 7.2. *The kinematics of the 1-1 machine implies that the tool trajectory is invariant with respect to* $T_{12,z}$, T_{23} *and* $T_{34,y}$.

The proof of Theorem 7.2 is analogous to that of Theorem 7.1.

Theorem 7.3. *The kinematics of the 0-2 machine implies that the tool trajectory is invariant with respect to* T_{12} *and* $T_{23,y}$.

The proof of Theorem 7.3 is analogous to that of Theorem 7.1.

7.2.2 Workpiece Setup and the Tool Trajectory

Changing the workpiece setup $\mathfrak{M}_S = \{T_{12}, r_a, r_b\}$ changes the coordinates of the cutter contact points, the orientations of the tool, and, consequently, the tool trajectory. Suppose the workpiece has been rotated clockwise by r_a around the $\mathbf{z}_1$-axis and by r_b around the $\mathbf{y}_1$-axis (Figure 7.1). The following theorems develop a closed-form representation of the tool trajectory for arbitrary r_a, r_b for the three basic types of the five-axis machines.

Theorem 7.4. *Let* r_a *and* r_b *be the rotation angles defining the orientation of the workpiece, and let* R_a *and* R_b *be the corresponding rotation matrices. The kinematics of the 2-0 machine is given by*

$$M = GB[b']\,(A[a']\,(W' + T_{12}) + T_{23}) + T_{34}^*, \tag{7.6}$$

where

$$\begin{aligned} W' &= R_b[r_b]R_a[r_a]W \\ &= \begin{bmatrix} \cos(r_b)\cos(r_a)x + \cos(r_b)\sin(r_a)y - \sin(r_b)z \\ -\sin(r_a)x + \cos(r_a)y \\ \sin(r_b)\cos(r_a)x + \sin(r_b)\sin(r_a)y + \cos(r_b)z \end{bmatrix}, \end{aligned}$$

$$I' = \begin{bmatrix} \cos(a - r_a)\cos(b)\cos(r_b) + \sin(b)\sin(r_b) \\ \sin(a - r_a)\cos(b) \\ \cos(a - r_a)\cos(b)\sin(r_b) - \sin(b)\cos(r_b) \end{bmatrix}.$$

Proof. The relationships between the tool orientation $I = (I_x, I_y, I_z)$ and the two rotary angles a and b are given by (see Sect. 2.4)

$$\begin{aligned} I_x &= \cos(a)\cos(b), \\ I_y &= \sin(a)\cos(b), \\ I_z &= -\sin(b). \end{aligned} \tag{7.7}$$

After the rotations, the tool orientation becomes

$$I' = R_b[r_a]R_a[r_b]I. \tag{7.8}$$

After some algebraic manipulation, (7.8) becomes

$$I' = \begin{bmatrix} \cos(a - r_a)\cos(b)\cos(r_b) + \sin(b)\sin(r_b) \\ \sin(a - r_a)\cos(b) \\ \cos(a - r_a)\cos(b)\sin(r_b) - \sin(b)\cos(r_b) \end{bmatrix} \tag{7.9}$$

Theorem 7.5. *Let r_a and r_b be the rotation angles defining the orientation of the workpiece, and let R_a and R_b be the corresponding rotation matrices. The kinematics of the 1-1 machine is given by*

$$M = GA[a'](W' + T_{12}) + T_{23} + B^{-1}[b']T^*_{34}, \tag{7.10}$$

where

$$I' = \begin{bmatrix} \cos(a + r_a)\sin(b)\cos(r_b) - \cos(b)\sin(r_b) \\ -\sin(b)sin(a + r_a) \\ \cos(a + r_a)sin(b)\sin(r_b) + \cos(b)\cos(r_b) \end{bmatrix}.$$

The proof is analogous to that of Theorem 7.4.

Theorem 7.6. *Let r_a and r_b be the rotation angles defining the orientation of the workpiece, and let R_a and R_b be the corresponding rotation matrices. The kinematics of the 0-2 machine is given by*

$$M = GW' + T_{12} + A^{-1}[a'](T_{23} + B^{-1}[b']T^*_{34}) \tag{7.11}$$

where

$$I' = \begin{bmatrix} (\cos(b)\cos(r_a) - \sin(a)\sin(b)\sin(r_a))\cos(r_b) - \cos(a)\sin(b)\sin(r_b) \\ -\cos(b)\sin(r_a) - \sin(a)\sin(b)\cos(r_a) \\ (\cos(b)\cos(r_a) - \sin(a)\sin(b)\sin(r_a))\sin(r_b) + \cos(a)\sin(b)\cos(r_b) \end{bmatrix}.$$

The proof is analogous to that of Theorem 7.4.

Finally, substituting the new rotation angles $\mathfrak{R}'_p = (a'_p, b'_p)$ obtained from I' and the new coordinates W' into (3.48) yields

$$\begin{aligned} W'_{p,p+1}(t) = \mathfrak{K}(\mathfrak{M}, t\mathfrak{R}'_{p+1} + (1-t)\mathfrak{R}'_p, \\ t\mathfrak{K}^{-1}(\mathfrak{M}, \mathfrak{R}'_{p+1}, W'_{p+1}) + (1-t)\mathfrak{K}^{-1}(\mathfrak{M}, \mathfrak{R}'_p, W'_p)), \end{aligned} \tag{7.12}$$

where $\mathfrak{K}$ corresponds to either (2.2) or (2.7) or (2.10).

7.3 Least-Squares Optimization and Dependent Variables

This section introduces a definition of the kinematics error in the least-squares sense. The optimal setup is defined as a set of optimization parameters minimizing the kinematics error. A system of non-linear equations to minimize the error is deduced and analyzed. The dependent optimization parameters are excluded from the minimization procedure. Finally, the minimal set of optimization parameters is presented for each machine type.

7.3.1 Least-Squares Optimization

Recall that $W^D_{p,p+1}(t) \equiv (x^D_{p,p+1}(t), y^D_{p,p+1}(t), z^D_{p,p+1}(t)) \in S(u,v)$ is defined as a curve between W_p and W_{p+1} extracted from the surface in such a way that it represents the desired tool trajectory between Π_p and Π_{p+1}. The error is represented as a deviation between $W^D_{p,p+1}(t)$ and $W_{p,p+1}(t) \equiv (x_{p,p+1}(t), y_{p,p+1}(t), z_{p,p+1}(t))$, namely,

$$
\begin{aligned}
\epsilon &= \sum_p \int_0^1 \left|W^D_{p,p+1}(t) - W_{p,p+1}(t)\right|^2 dt, \\
&= \sum_p \int_0^1 \Big[\left(x^D_{p,p+1}(t) - x_{p,p+1}(t)\right)^2 + \left(y^D_{p,p+1}(t) - y_{p,p+1}(t)\right)^2 \\
&\qquad + \left(z^D_{p,p+1}(t) - z_{p,p+1}(t)\right)^2\Big]\, dt,
\end{aligned}
\tag{7.13}
$$

where

$$
\begin{aligned}
W_{p,p+1}(t) = R_a^{-1}[r_a]R_b^{-1}[r_b]\mathfrak{K}(\mathfrak{M}, t\mathfrak{R}'_{p+1} + (1-t)\mathfrak{R}'_p,\\
t\mathfrak{K}^{-1}(\mathfrak{M}, \mathfrak{R}'_{p+1}, R_b[r_b]R_a[r_a]W_{p+1})\\
+ (1-t)\mathfrak{K}^{-1}(\mathfrak{M}, \mathfrak{R}'_p, R_b[r_b]R_a[r_a]W_p)).
\end{aligned}
$$

Consider now the least-squares minimization problem described by (7.2). Differentiating with regard to the optimization parameters yields

$$
\frac{\partial \epsilon}{\partial \nu_p} = \sum_p \int_0^1 \left|W^D_{p,p+1}(t) - W_{p,p+1}(t)\right| \frac{\partial W_{p,p+1}(t)}{\partial \nu_p} dt = 0,
\tag{7.14}
$$

where $\nu = (\nu_1, \ldots, \nu_n)$ is a vector comprising the optimization parameters. The number of parameters n is at most six, but if the optimization is only performed with respect to the machine setup, the number of the parameters will vary with the type of machine.

7.3.2 Dependent Variables

Let us now analyze dependent variables. As an example, consider the 2-0 machine. Observe that $T_{34} - T_4$ has been replaced by one variable T_{34}^* (see Sect. 2.4). Consequently, if $T_{34}^* = T_{34,o}^*$ is a component of the solution of the transformed optimization problem then any T_{34} and T_4 that satisfy $T_{34} - T_4 = T_{34,o}^*$ are components of the solutions of the original optimization problem. In terms of the cost function, the dependence is expressed as follows: ν_1 and ν_2 are dependent if $\epsilon(\nu_1, \nu_2, \nu_3, \ldots, \nu_n) \equiv \hat{\epsilon}(c_1\nu_1 + c_2\nu_2, \nu_3, \ldots, \nu_n)$, where c_1 and c_2 are constants. In this case $\frac{\partial \hat{\epsilon}}{\partial v_1} = 0$ and $\frac{\partial \hat{\epsilon}}{\partial v_2} = 0$ produce the same equations. Therefore, $c_1\nu_1 + c_2\nu_2$ can be replaced by one variable. When the optimal solution to the transformed problem is found, either ν_1 or ν_2 is considered a free variable.

The definition of dependence can be easily generalized to the case of n dependent variables. It will now be established that (7.13) for the kinematics in (2.2) has an additional, less obvious, pair of dependent variables.

Theorem 7.7. *$T_{12,z}$ and $T_{23,z}$ are linearly dependent for the 2-0 machine.*

Proof. Consider a tool trajectory given by $W_{p,p+1}(t) = \hat{\mathfrak{K}}(tB^{-1}B_{p+1}\hat{\mathfrak{K}}_{p+1}^{-1} + (1-t)B^{-1}B_p\hat{\mathfrak{K}}_p^{-1})$, where $\hat{\mathfrak{K}}_p^{-1} = A_p(W_p + T_{12}) + T_{23}$ (see (7.4)). Since A_p and A_{p+1} are rotations around the **z**-axis, they do not change the z-component of T_{12}. Therefore, $(A_pT_{12} + T_{23})_z = (A_{p+1}T_{12} + T_{23})_z = T_{12,z} + T_{23,z}$. Hence, $(A_pT_{12} + T_{23})_z$ and $(A_{p+1}T_{12} + T_{23})_z$ can be replaced by one variable. Furthermore, since $\hat{\mathfrak{K}}(S) = A^{-1}(S - T_{23}) - T_{12}$, $(A^{-1}T_{23} + T_{12})_z = T_{23,z} + T_{12,z}$ can be replaced by the same variable. Consequently, $T_{12,z}$ and $T_{23,z}$ are dependent.

Remark 7.1. The dependency of $T_{12,z}$ and $T_{23,z}$ is illustrated by Fig. 7.3 (see also Fig. 2.4 for comparison). In words, since $T_{12,z} + T_{23,z} = c$ (c is the distance between the center of the workpiece coordinate system O_1 and the center of the coordinate system of the second rotary part O_3), the optimal workpiece position relative to the second rotary part is fixed. However, the height of the mounting table can be reduced or increased along the $\mathbf{z}_2$-axis (the O_2 coordinate system).

Consider now the 1-1 and the 0-2 machines.

Theorem 7.8. *T_{34} and T_4 are linearly dependent for the 1-1 machine.*

The proof is analogous to that of Theorem 7.7.

Theorem 7.9. *$T_{23,x}$ and $T_{34,x}$ are linearly dependent for the 0-2 machine.*

The proof is analogous to that of Theorem 7.7.

The results presented in this section are summarized in Table 7.1.

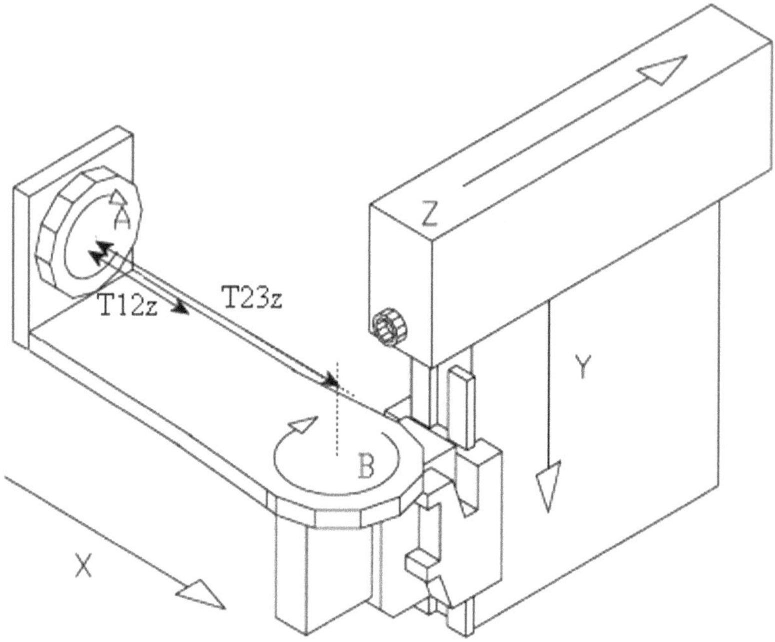

Fig. 7.3. Dependent parameters $T_{12,z} + T_{23,z} = c$ of the 2-0 machine

Table 7.1. Minimal sets of optimization parameters

Machine type	Optimization parameters		
	Workpiece setup	Tool length	Machine settings
The 2-0 machine	$r_a, r_b, T_{12,x}, T_{12,y}, T_{12,z}$	none	$T_{23,x}$
The 1-1 machine	$r_a, r_b, T_{12,x}, T_{12,y}$	$T_{4,z}$	$T_{34,x}$
The 0-2 machine	r_a, r_b	$T_{4,z}$	$T_{23,x}, T_{23,y}, T_{34,y}$

7.4 Examples and Discussion

7.4.1 Numerical Method

This section describes a numerical solution of system (7.14) by a standard Newton-Raphson procedure. Consider a system $F(v) = 0$, where F is a vector function deduced from (7.14). A number of numerical methods can be used to solve this system of nonlinear equations [1]. Among the simplest is the standard Newton method given by

$$\nu^{\text{new}} = \nu^{\text{old}} - J^{-1} F(\nu^{\text{old}}), \tag{7.15}$$

where J is the Jacobi matrix. It is well known that the performance of Newton's method depends on the initial approximation. The process may not converge, and the global optimum may not be obtained in the case of multiple optima. To counter these limitations, Newton's method is executed several times with a different initial approximation in each run. Starting points are arranged in a regular grid pattern or sampled from a uniform distribution over feasible parameter space. The required derivatives of the objective function with respect to the rotation angles are obtained with the chain rule, as follows:

$$\frac{\partial\epsilon(r_a, r_b)}{\partial r_L} = \frac{\partial\epsilon}{\partial r_L} + \frac{\partial\epsilon}{\partial a'_p}\frac{\partial a'_p}{\partial r_L} + \frac{\partial\epsilon}{\partial b'_p}\frac{\partial b'_p}{\partial r_L} + \frac{\partial\epsilon}{\partial a'_{p+1}}\frac{\partial a'_{p+1}}{\partial r_L} + \frac{\partial\epsilon}{\partial b'_{p+1}}\frac{\partial b'_{p+1}}{\partial r_L}, \quad (7.16)$$

where $L = a$ or $L = b$. a' and b' are machine rotation angles which are function of r_a and r_b (see Theorems 7.4-7.6). The next section demonstrates the performance of this simple approach with several experiments.

7.4.2 Examples

To demonstrate the performance of the proposed procedure, the optimization is applied to two test surfaces, a sweep surface [4, 5] (Figure 7.4(a)) and a two-bell surface (Figure 7.4(b)). The sweep surface is given by

$$\begin{aligned} x &= 100u - 50, \\ y &= 100v - 50, \\ z &= -40(v - 0.8)^2 - 50\sin(3u - 0.3)\sin(u - 0.5)\sin(u - 0.8) - 5. \end{aligned} \quad (7.17)$$

The two-bell surface is given by

$$\begin{aligned} x &= 100u - 50, \\ y &= 100v - 50, \\ z &= 400v(1 - v)(3.55u - 14.8u^2 + 21.15u^3 - 9.9u^4) - 28. \end{aligned} \quad (7.18)$$

where $0 \leq u, v \leq 1$.

Let $W_p = S(u_p, v_p)$ and $W_{p+1} = S(u_{p+1}, v_{p+1})$ be two successive CC points. The desired curve between the two CC points is extracted from the surface as follows:

$$W^D_{p,p+1}(t) = S((1 - t)u_p + tu_{p+1}, (1 - t)v_p + tv_{p+1}), \quad (7.19)$$

where t is a fictitious time coordinate along the curve ($0 \leq t \leq 1$). The mean error is an approximation of (7.13), given by

$$\bar{\epsilon} = \frac{1}{N_{pt}} \sum_p \sum_t \left|W^D_{p,p+1}(t) - W_{p,p+1}(t)\right|^2, \quad (7.20)$$

where N_{pt} is the total number of sampled points.

The performance of the optimization is presented by Table 7.2 (the sweep surface) and Table 7.3 (the two-bell surface).

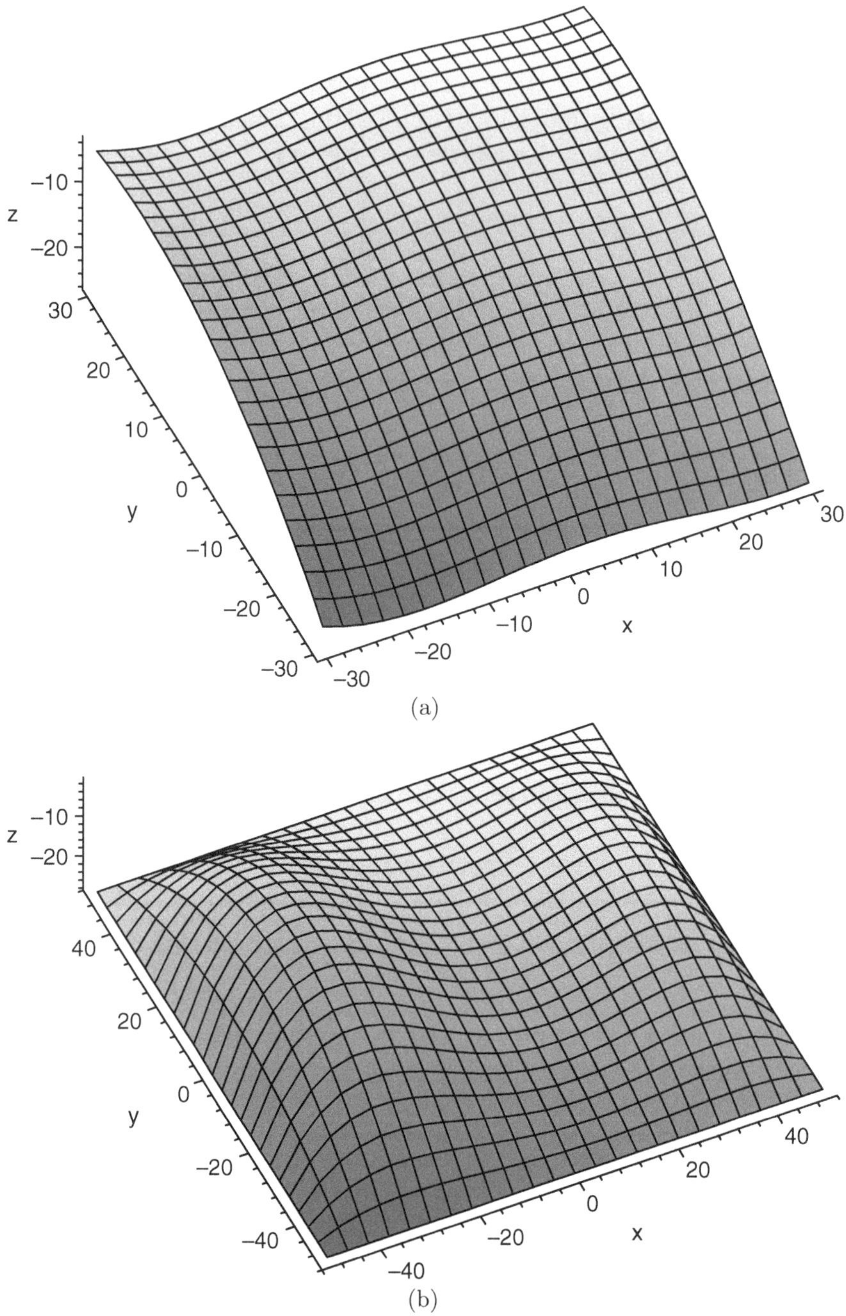

Fig. 7.4. Test surfaces: (**a**) sweep surface, (**b**) two-bell surface

Table 7.2. Performance of the optimization method, the sweep surface

Machine type	The mean error (mm) / reduction			
	Before optimization	Optimization wrt. the workpiece setup	Optimization wrt. the workpiece setup and the tool length	Optimization wrt. the entire set
The 2-0 machine	0.5730	0.0179 / 96.88%	-	0.0176 / 96.92%
The 1-1 machine	0.5785	0.0167 / 97.12%	0.0146 / 97.47%	0.0145 / 97.49%
The 0-2 machine	0.0522	0.0396 / 24.14%	0.0273 / 47.78%	0.0160 / 69.34%

Table 7.3. Performance of the optimization method, the two-bell surface

Machine type	The mean error (mm) / reduction			
	Before optimization	Optimization wrt. the workpiece setup	Optimization wrt. the workpiece setup and the tool length	Optimization wrt. the entire set
The 2-0 machine	0.5408	0.1118 / 79.33%	-	0.1102 / 79.62%
The 1-1 machine	0.6208	0.5064 / 18.43%	0.2136 / 65.59%	0.2133 / 65.64%
The 0-2 machine	0.5564	0.5403 / 2.90%	0.1572 / 71.75%	0.0454 / 91.83%

The desired and actual tool trajectory obtained for the sweep surface is depicted in Figs. 7.5-7.7 along with the corresponding error. The tool trajectories are depicted in the original workpiece coordinates, i.e., each tool trajectory, beginning from the optimal setup, is transformed back to workpiece coordinates for comparison. The proposed optimization is compared with the results obtained by traditional positioning of the workpiece, in which the origin of the workpiece coordinates is aligned with the center of the mounting table, and the **z**-axis is perpendicular to the mounting table. The results reveal that the error can be reduced by as much as 97%. Table 7.3 shows a similar accuracy increase achieved for the geometrically complex two-bell surface.

The sweep surface produced by MAHO600E (a 2-0 machine, Fig. 2.1) with and without optimization is shown in Fig. 7.8. The surface in Fig. 7.8(d) was produced with a setup optimized with regard to r_a and r_b. The optimization leads to an error reduction of about 81.45%. Note that in order to better visualize the errors, a very small number of CC points was used (102 points).

Another measure to evaluate the performance of optimization is the number of CC points required to maintain the kinematics error within a prescribed tolerance. This estimate is in particular important for high speed milling when an increase in the number of the points leads to a substantial increase in the machining time.

Table 4 displays the number of CC points required to cut the workpiece to within a prescribed maximum machining error. The optimal setup applied to both surfaces, reduces the number of CC points by 68% and 30%, respectively. Note that although the reduction achieved for the two-bell surface is not particularly large, the optimization is still appropriate since it requires nothing of the operator but changing the initial setup. This is, a simple, nearly zero cost, operation.

Table 7.4. Performance of the optimization measured by the required number of CC points

	Number of CC points / reduction	
Surface	Before optimization	Optimization wrt. r_a and r_b
Sweep surface	3900	1233 / 68.4%
Two-bell surface	7925	5544 / 30.0%

2-0 machine. The maximum allowed machining error = 0.01 mm.

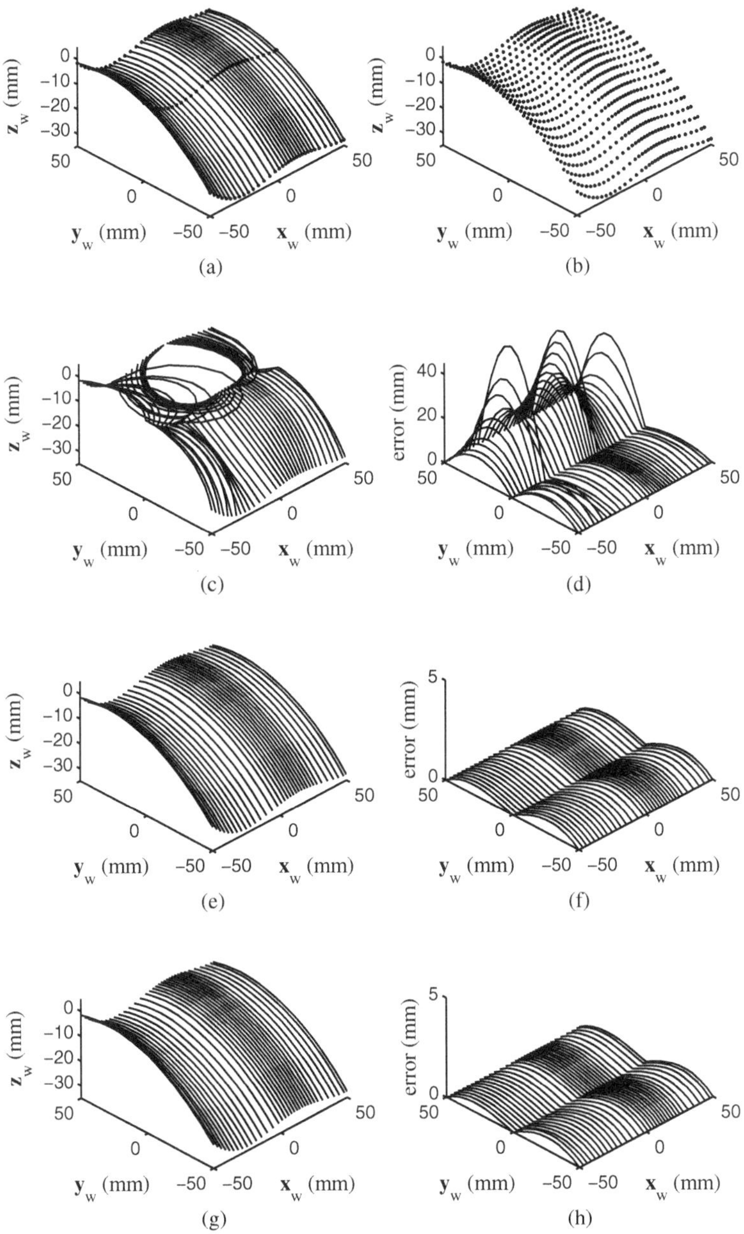

Fig. 7.5. (**a**) CC points on the desired tool trajectory, (**b**) the extracted tool trajectory (sampling points), (**c**) the actual tool trajectory on the 2-0 machine before optimization and (**d**) the error, (**e**) the tool trajectory with an optimal setup r_a, r_b, T_{12} and (**f**) the error, (**g**) the tool trajectory with an optimal setup r_a, r_b, T_{12}, T_{23} and (**h**) the error

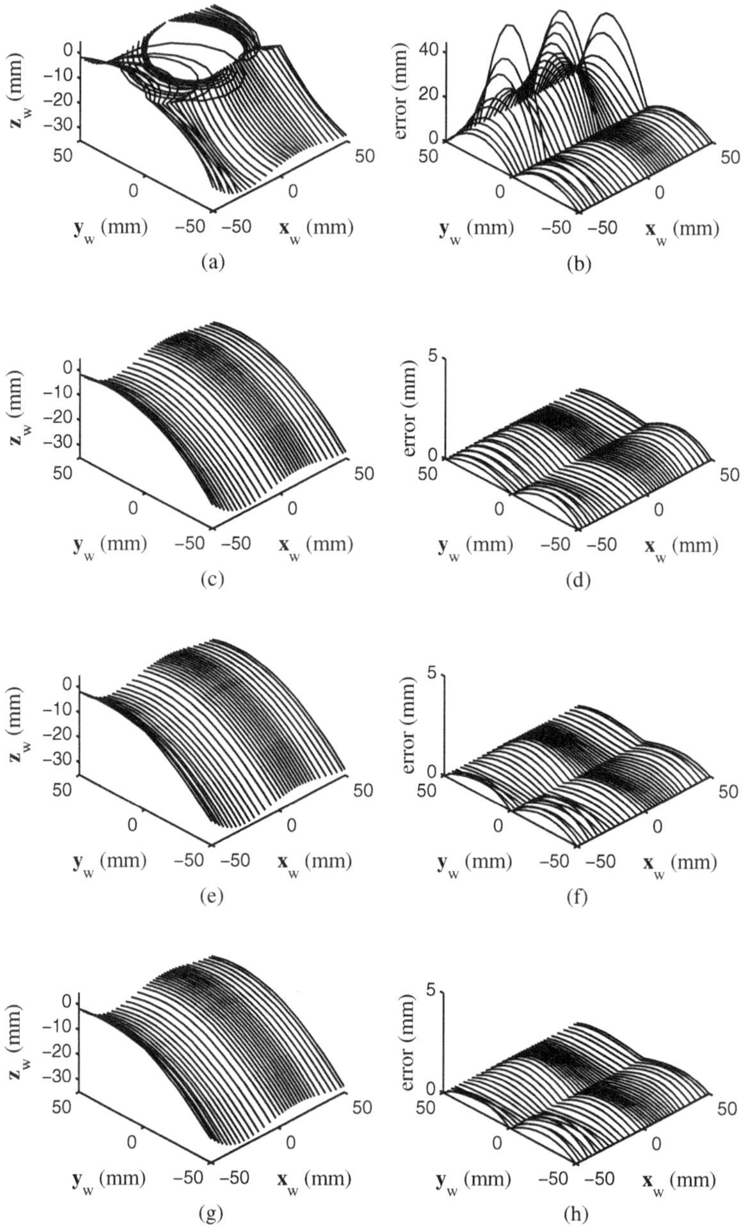

Fig. 7.6. (**a**) the actual tool trajectory on the 1-1 machine before optimization and (**b**) the error, (**c**) the tool trajectory with an optimal setup r_a, r_b, T_{12} and (**d**) the error, (**e**) the tool trajectory with an optimal setup r_a, r_b, T_{12}, T_4 and (**f**) the error, (**g**) the tool trajectory with an optimal setup r_a, r_b, T_{12}, T_{34}, T_4 and (**h**) the error

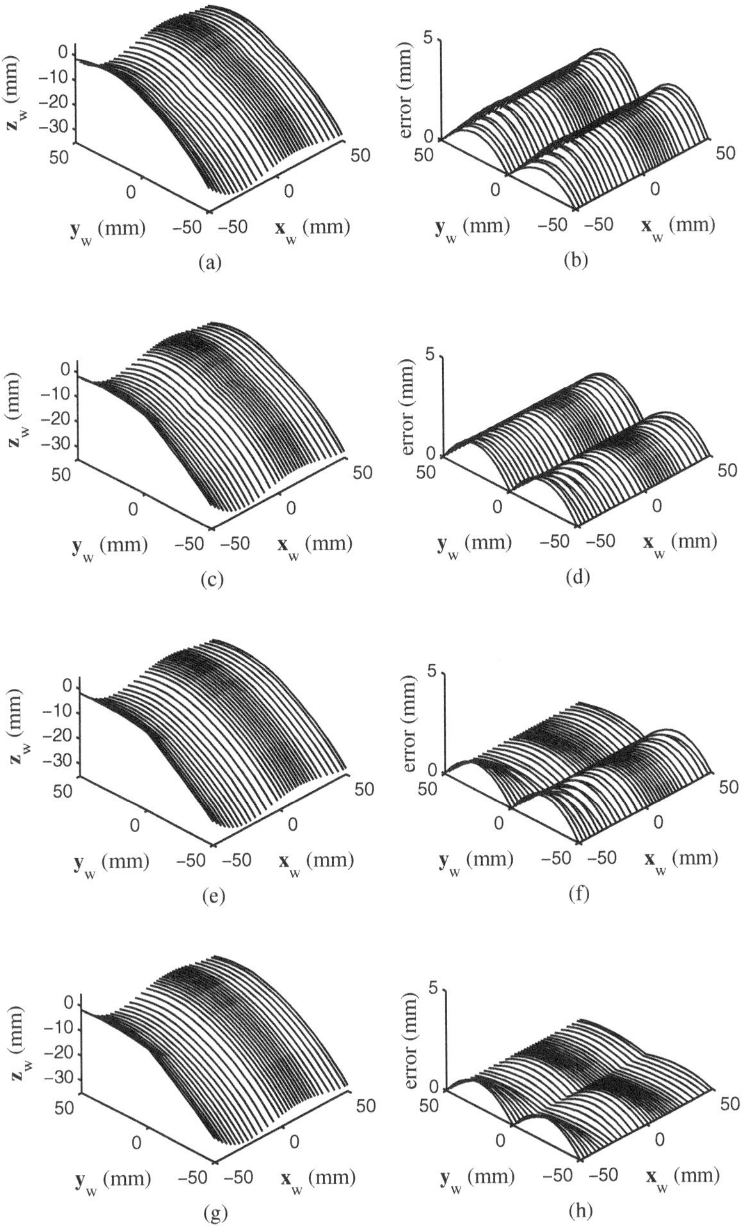

Fig. 7.7. (**a**) the actual tool trajectory on the 0-2 machine before optimization and (**b**) the error, (**c**) the tool trajectory with an optimal setup r_a, r_b and (**d**) the error, (**e**) the tool trajectory with an optimal setup r_a, r_b, T_4 and (**f**) the error, (**g**) the tool trajectory with an optimal setup r_a, r_b, T_{23}, T_{34}, T_4 and (**h**) the error

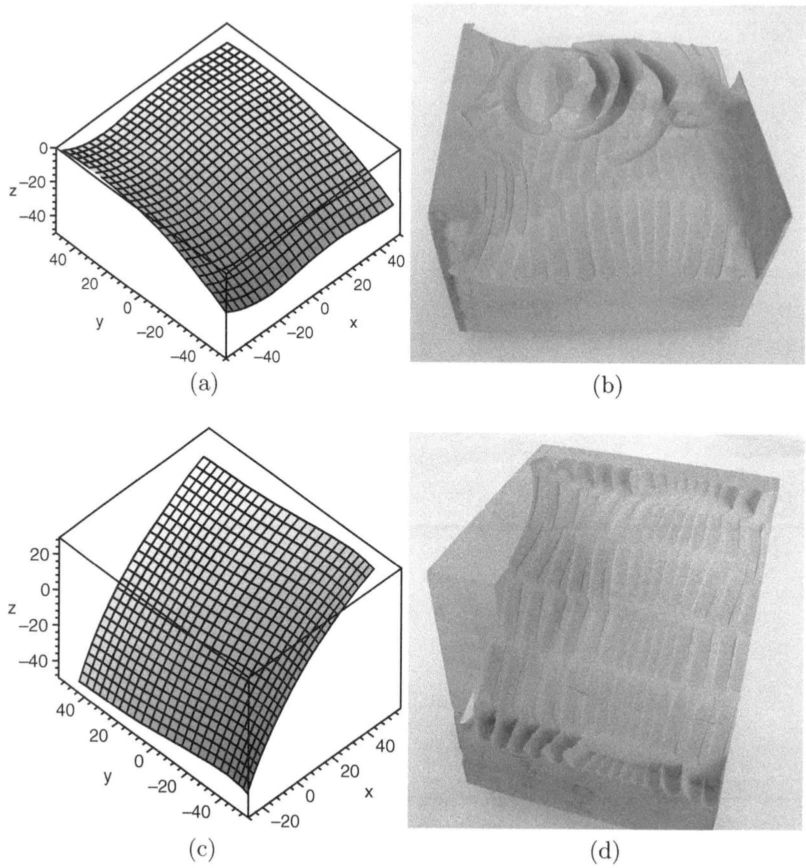

Fig. 7.8. The sweep surface machined with different setups. The workpiece on the top (**a**) and (**b**) has been cut with a conventional setup ($r_a = 0$, $r_b = 0$). The workpiece at the bottom (**c**) and (**d**) has been cut with $r_a = 1.6008$ radian and $r_b = 0.5236$ radian (the optimal setup with regard to r_a and r_b)

References

[1] Bertsekas, D. P. 1995. *Nonlinear Programming.* Athena Scientific, Belmont, MA, USA.
[2] Bohez, E. L. J. 2002. Five-axis milling machine tool kinematic chain design and analysis. *International Journal of Machine Tools and Manufacture*, 42(4):505–520.
[3] Jung, Y. H., Lee, D. W., Kim, J. S., and Mok, H. S. 2002. NC postprocessor for 5-axis milling machine of table-rotating/tilting type. *Journal of Materials Processing Technology*, 130-131:641–646.
[4] Piegl, L. and Tiller, W. 1995. *The NURBS book.* Springer-Verlag, London, UK.

[5] Rogers, D. F. 2001. *An introduction to NURBS: with historical perspective.* Morgan Kaufmann Publishers Inc., San Francisco, CA, USA.

[6] Sijie, Y., Yunfei, Z., Fangyu, P., and Xide, L. 2004. Research on the localisation of the workpieces with large sculptured surfaces in NC machining. *The International Journal of Advanced Manufacturing Technology*, 23(5-6):429–435.

[7] Tutunea-Fatan, O. R. and Feng, H.-Y. 2004. Configuration analysis of five-axis machine tools using a generic kinematic model. *International Journal of Machine Tools and Manufacture*, 44(11):1235–1243.

[8] Warkentin, A., Hoskins, P., Ismail, F., and Bedi, S. 2001. Computer-aided 5-axis machining. In *Systems Techniques and Computational Methods*, volume 1 of *Computer-Aided Design, Engineering, and Manufacturing: Systems Techniques and Applications*, chapter 3, pages 3001–3034. CRC Press, Inc., Boca Raton, FL, USA.

Index

FSC
www.fsc.org
MIX
Papier | Fördert
gute Waldnutzung
FSC® C083411

Kolibri 360 GmbH
Ferdinand-Jühlke-Straße 7
99095 Erfurt, Deutschland
produktsicherheit@kolibri360.de